Endless Forms Most Beautiful

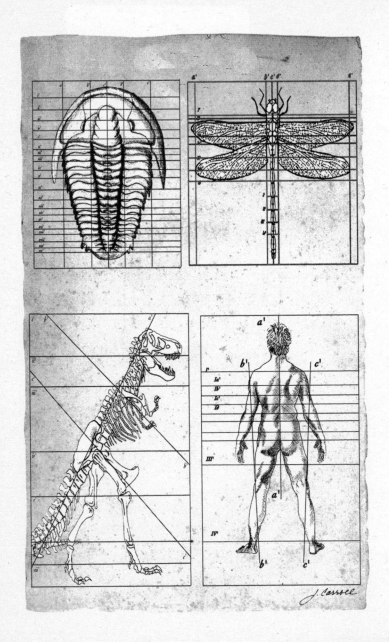

J. Carroll

Endless Forms Most Beautiful

The New Science of Evo Devo

AND THE MAKING OF THE ANIMAL KINGDOM

Sean B. Carroll

WITH ILLUSTRATIONS BY

Jamie W. Carroll ∗ *Josh P. Klaiss* ∗ *Leanne M. Olds*

Quercus

First published in Great Britain in 2006 by Weidenfeld & Nicolson

This edition published in 2011 by
Quercus
21 Bloomsbury Square
London
WC1A 2NS

A CIP catalogue record for this book is available
from the British Library

ISBN 978 1 84916 048 3

Design by Lovedog Studio

Printed and bound in Great Britain by Clays Ltd, St Ives plc

For Jamie, Will, Patrick,

Chris, and Josh

Contents

Revolution #3

You say you want a revolution
Well, you know
we all want to change the world.
You tell me that it's evolution,
Well, you know
we all want to change the world . . .
You say you got a real solution
Well, you know
we'd all love to see the plan. . . .

— John Lennon and Paul McCartney
"Revolution 1" (1968)

THE PHYSICIST AND NOBEL laureate Jean Perrin once stated that the key to any scientific advance is to be able "to explain the complex visible by some simple invisible." The two greatest revolutions in biology, those in evolution and genetics, were driven by such insights. Darwin explained the parade of species in the fossil record and the diversity of living organisms as products of natural selection over eons of time. Molecular biology explained how the basis of heredity in all species is encoded in molecules of DNA made of just four basic constituents. As powerful as these insights were, in terms of explaining the origin of complex visible *forms*, from the bodies of ancient trilobites to the beaks of Galapagos finches, they were incomplete. Neither natural selection nor DNA directly explains *how* individual forms are made or how they evolved.

The key to understanding form is *development*, the process through which a single-celled egg gives rise to a complex, multi-billion-celled animal. This amazing spectacle stood as one of the great unsolved mysteries of biology for nearly two centuries. And development is intimately connected to evolution because it is through changes in embryos that changes in form arise. Over the past two decades, a new revolution has unfolded in biology. Advances in developmental biology and evolutionary developmental biology (dubbed "Evo Devo") have revealed a great deal about the invisible genes and some simple rules that shape animal form and evolution. Much of what we have learned has been so stunning and unexpected that it has profoundly reshaped our picture of how evolution works. Not a single biologist, for example, ever anticipated that the same genes that control the making of an insect's body and organs also control the making of our bodies.

This book tells the story of this new revolution and its insights into how the animal kingdom has evolved. My goal is to reveal a vivid picture of the process of making animals and how various kinds of changes in that process have molded the different kinds of animals we know today and those from the fossil record.

I wrote the book with several kinds of readers in mind. First, for anyone interested in nature and natural history who takes delight in animals of the rain forest, reef, savannah, or fossil beds, there will be much said about the making and evolution of some of the most fascinating animals of the past and present. Second, for physical scientists, engineers, computer scientists, and others interested in the origins of complexity, this book tells the story of the enormous diversity that has been created from combining a small number of common ingredients. Third, for students and educators, I firmly believe that the new insights from Evo Devo bring the evolutionary process alive and offer a more gripping and illuminating picture of evolution than has typically been taught and discussed. And fourth, for anyone who may ponder the question "Where did I come from?," this book is also about our history, both the journey we have all made from egg to adult, and the long trek from the origin of animals to the very recent origin of our species.

Endless Forms Most Beautiful

Drawing by Christopher Herr, age ten (Eagle School, Madison, Wisconsin)

Introduction

Butterflies, Zebras, and Embryos

> Well she's walking through the clouds
> With a circus mind that's running round
> Butterflies and zebras
> And moonbeams and fairy tales,
> That's all she ever thinks about. . . .

> —Jimi Hendrix,
> "Little Wing" (1967)

ON A RECENT VISIT to my kids' elementary school, I was enjoying the student art that decorated the hallways. Among the landscapes and portraits were many depictions of animals. I couldn't help noting that of the thousands of species to choose from, the most frequently drawn mammal was the zebra. And the most represented animal of any kind was the butterfly. We live in Wisconsin and, this being the middle of winter, the kids were not drawing what they saw out the window. So, why all the butterflies and zebras?

I am certain that these pieces of art reflect the children's deep connection to animal forms—their shapes, patterns, and colors. We all feel that connection. That's why we visit zoos to see exotic animals, flock to the new phenomenon of butterfly aviaries, go to aquaria, and spend billions on our animal companions—dogs, cats, birds, and even fish. We most often choose our favorite breeds and species on aesthetic grounds. Yet, we are also mesmerized, and sometimes terrified, by the more extreme animal forms: giant squids, carnivorous dinosaurs, or bird-eating spiders.

The same attraction to and fascination with animal forms has motivated the greatest naturalists for centuries. In cold, gray, damp pre-Victorian England, young Charles Darwin read Alexander von Humboldt's *Personal Narrative*, a 2000-page account of his voyage to and around South America. Darwin was so consumed that he later claimed that all he thought, spoke, or dreamt about were schemes to get to see the sights of the Tropics that Humboldt described. He leaped at the chance when the opportunity to sail on the *Beagle* arose in 1831. Darwin later wrote to Humboldt, "My whole course of life is due to having read and re-read as a youth, this personal narrative." Two other Englishmen, Henry Walter Bates, a twenty-two-year-old office clerk and avid bug collector, and his self-taught naturalist friend, Alfred Russel Wallace, also dreamt of travel abroad to collect new species. Upon reading an American's account of a journey to Brazil, Bates and Wallace immediately decided to head there (in 1848). Darwin's voyage lasted five years, Bates remained in the Tropics for eleven years, and Wallace spent fourteen years over the course of two journeys. These dreamers would, based upon the thousands of species that they saw and collected, launch the first revolution in biology.

There must be something about living in northern climates that inspires dreams of the Tropics. I grew up in Toledo, Ohio, surrounded by city parks and farm fields, near the shores of the less than bountiful Lake Erie. My dreams of paradise were fed by magazines and the TV show *Animal Kingdom* (broadcast in black and white). Decades later, I have been lucky enough to see the animals of the African savannah, the jungles of Central America, and the barrier reefs of Australia and Belize (as a tourist, not as a courageous explorer—trust me). And they are even more awe-inspiring than I had imagined.

On the open grasslands of Kenya, herds of zebras and elephants graze while solitary giraffes, ostriches, and cheetahs stroll by. Striped horses, gigantic gray mammals with six-foot-long noses, and spotted cats that can outrun a Jeep? If these creatures did not exist, they would be almost too incredible to believe.

In the rain forests, the richness is generally in smaller creatures. In the dappled light created by gaps in the canopy, brightly painted butterflies such as the red and yellow *Heliconius* or the sparkling metallic blue *Morpho* dance. In the litter below, red-and-turquoise-splotched poison arrow frogs call and vivid green leaf-cutter ants go about their vast harvesting projects. The big predators come out at night. I shall never forget the thrill of meeting a six-foot-long, deadly fer-de-lance snake in the pitch darkness and absolute quiet of a jungle in Belize, in a place well populated with jaguars (we saw only fresh tracks, but that was close enough!).

The sea holds even more strange and wonderful forms. Plunge into the shallow waters off an Australian coral island and the variety of fish, corals, and shelly creatures will literally hit you in the face. Neon colors, bodies of all shapes and sizes, fantastic geometrical designs are everywhere, and occasionally there's a glimpse of a giant sea turtle, an octopus, or a darting shark.

The great variety in the size, shape, organization, and color of animal bodies raises deep questions about the origins of animal forms. How are individual forms generated? And how have such diverse forms evolved? These are very old questions in biology, which date back to the time of Darwin, Wallace, and Bates and before, but only very recently have deep answers been discovered, many of them so surprising and profound that they have revolutionized our views on the making of the animal world and our place in it. The initial inspiration for this story is the attraction we all share to animal form, but my aim is to expand that wonder and fascination to *how* form is created—that is, to our new understanding of the biological processes that generate pattern and diversity in animal design. Underlying the many visible elements of animal form are remarkable processes, beautiful in their own right in the way that they transform a tiny, single cell into a large, complex, highly organized, and patterned creature, and over time, have forged a kingdom of millions of individual designs.

Embryos and Evolution

The first approach naturalists took to dealing with the great variety of animals was to sort them into groups, such as vertebrates (including fish, amphibians, reptiles, birds, and mammals) and arthropods (insects, crustaceans, arachnids, and more), but between and within these groups there are many differences. What makes a fish different from a salamander? Or an insect from a spider? On a finer scale, clearly a leopard is a cat, but what makes it different from a domestic tabby? And closer to home, what makes us different from our chimpanzee cousins?

The key to answering such questions is to realize that every animal form is the product of two processes—development from an egg and evolution from its ancestors. To understand the origins of the multitude of animal forms, we must understand these two processes and their intimate relationship to each other. Simply put, development is the process that transforms an egg into a growing embryo and eventually an adult form. The evolution of form occurs through changes in development.

Both processes are breathtaking. Consider that the development of an entire complex creature begins with a single cell—the fertilized egg. In a matter of just a day (a fly maggot), a few weeks (a mouse), or several months (ourselves), an egg grows into millions, billions, or, in the case of humans, perhaps 10 trillion cells formed into organs, tissues, and parts of the body. There are few, if any, phenomena in nature that inspire our wonder and awe as much as the transformation from egg to embryo to the complete animal. One of the great figures in all of biology, Darwin's close ally Thomas H. Huxley, remarked:

> The student of Nature wonders the more and is astonished the less, the more conversant he becomes with her operations; but of all the perennial miracles she offers to his inspection, perhaps the most worthy of admiration is the development of a plant or of an animal from its embryo.

Aphorisms and Reflections (1907)

The intimate connection between development and evolution has long been appreciated in biology. Both Darwin, in *The Origin of Species* (1859) and *The Descent of Man* (1871), and Huxley in his short masterpiece, *Evidence as to Man's Place in Nature* (1863), leaned heavily on the facts of embryology (as they were in the mid-nineteenth century) to connect man to the animal kingdom and for indisputable evidence of evolution. Darwin asked his reader to consider how slight changes, introduced at different points in the process and in different parts of the body, over the course of many thousands or a million generations, spanning perhaps tens of thousands to a few million years, can produce different forms that are adapted to different circumstances and that possess unique capabilities. That is evolution in a nutshell.

For Huxley, the nub of the argument was simple: we may marvel at the process of an egg becoming an adult, but we accept it as an everyday fact. It is merely then a lack of imagination to fail to grasp how changes in this process that are assimilated over long periods of time, far longer than the span of human experience, shape life's diversity. Evolution is as natural as development.

> As a natural process, of the same character as the development of a tree from its seed, or of a fowl from its egg, evolution excludes creation and all other kinds of supernatural intervention.
>
> *Aphorisms and Reflections* (1907)

While Darwin and Huxley were right about development as key to evolution, for more than one hundred years after their chief works, virtually no progress was made in understanding the mysteries of development. The puzzle of how a simple egg gives rise to a complete individual stood as one of the most elusive questions in all of biology. Many thought that development was hopelessly complex and would involve entirely different explanations for different types of animals. So frustrating was the enterprise that the study of embryology, heredity, and evolution, once intertwined at the core of biological thought a century ago, fractured into separate fields as each sought to define its own principles.

Because embryology was stalled for so long, it played no part in the so-called Modern Synthesis of evolutionary thought that emerged in the 1930s and 1940s. In the decades after Darwin, biologists struggled to understand the mechanisms of evolution. At the time of *The Origin of Species*, the mechanism for the inheritance of traits was not known. Gregor Mendel's work was rediscovered decades later and genetics did not prosper until well into the 1900s. Different kinds of biologists were approaching evolution at dramatically different scales. Paleontology focused on the largest time scales, the fossil record, and the evolution of higher taxa. Systematists were concerned with the nature of species and the process of speciation. Geneticists generally studied variation in traits in just a few species. These disciplines were disconnected and sometimes hostile over which offered the most worthwhile insights into evolutionary biology. Harmony was gradually approached through an integration of evolutionary viewpoints at different levels. Julian Huxley's book *Evolution: The Modern Synthesis* (1942) signaled this union and the general acceptance of two main ideas. First, that gradual evolution can be explained by small genetic changes that produce variation which is acted upon by natural selection. Second, that evolution at higher taxonomic levels and of greater magnitude can be explained by these same gradual evolutionary processes sustained over longer periods.

The Modern Synthesis established much of the foundation for how evolutionary biology has been discussed and taught for the past sixty years. However, despite the monikers of "Modern" and "Synthesis," it was incomplete. At the time of its formulation and until recently, we could say that forms do change, and that natural selection is a force, but we could say nothing about *how* forms change, about the visible drama of evolution as depicted, for example, in the fossil record. The Synthesis treated embryology as a "black box" that somehow transformed genetic information into three-dimensional, functional animals.

The stalemate continued for several decades. Embryology was preoccupied with phenomena that could be studied by manipulating the eggs and embryos of a few species, and the evolutionary framework faded from embryology's view. Evolutionary biology was studying genetic variation in populations, ignorant of the relationship between genes and form. Perhaps even worse, the perception of evolutionary biology in some circles was that it had become relegated to dusty museums.

Such was the setting in the 1970s when voices for the reunion of embryology and evolutionary biology made themselves heard. Most notable was that of Stephen Jay Gould, whose book *Ontogeny and Phylogeny* revived discussion of the ways in which the modification of development may influence evolution. Gould had also stirred up evolutionary biology when, with Niles Eldredge, he took a fresh look at the patterns of the fossil record and forwarded the idea of *punctuated equilibria*—that evolution was marked by long periods of stasis (equilibria) interrupted by brief intervals of rapid change (punctuation).

Gould's book and his many subsequent writings reexamined the "big picture" in evolutionary biology and underscored the major questions that remained unsolved. He planted seeds in more than a few impressionable young scientists, myself included.

To me, and others who had been weaned on the emerging successes of molecular biology in explaining how genes work, the situations in embryology and in evolutionary biology were both unsatisfying, but they presented enormous potential opportunities. Our lack of embryological knowledge seemed to turn much of the discussion in evolutionary biology about the evolution of form into futile exercises in speculation. How could we make progress on questions involving the evolution of form without a scientific understanding of how form is generated in the first place? Population genetics had succeeded in establishing the principle that evolution is due to changes in genes, but this was a principle without an example. No gene that affected the form and evolution of any animal had been characterized. New insights in evolution would require breakthroughs in embryology.

The Evo Devo Revolution

Everyone knew that genes must be at the center of the mysteries of both development and evolution. Zebras look like zebras, butterflies look like butterflies, and we look like we do because of the genes we carry. The problem was that there were very few clues as to which genes mattered for the development of any animal.

The long drought in embryology was eventually broken by a few brilliant geneticists who, while working with the fruit fly, the workhorse of genetics for the past eighty years, devised schemes to find the genes that controlled fly development. The discovery of these genes and their study in the 1980s gave birth to an exciting new vista on development and revealed a logic and order underlying the generation of animal form.

Almost immediately after the first sets of fruit fly genes were characterized came a bombshell that triggered a new revolution in evolutionary biology. For more than a century, biologists had assumed that different types of animals were genetically constructed in completely different ways. The greater the disparity in animal form, the less (if anything) the development of two animals would have in common at the level of their genes. One of the architects of the Modern Synthesis, Ernst Mayr, had written that "the search for homologous genes is quite futile except in very close relatives." But contrary to the expectations of *any* biologist, most of the genes first identified as governing major aspects of fruit fly body organization were found to have exact counterparts that did the same thing in most animals, including ourselves. This discovery was followed by the revelation that the development of various body parts such as eyes, limbs, and hearts, vastly different in structure among animals and long thought to have evolved in entirely different ways, was also governed by the same genes in different animals. The comparison of developmental genes between species became a new discipline at the interface of embryology and evolutionary biology—evolutionary developmental biology, or "Evo Devo" for short.

The first shots in the Evo Devo revolution revealed that despite their great differences in appearance and physiology, all complex animals—flies and flycatchers, dinosaurs and trilobites, butterflies and zebras and humans—share a common "tool kit" of "master" genes that govern the formation and patterning of their bodies and body parts. I'll describe the discovery of this tool kit and the remarkable properties of these genes in detail in chapter 3. The important point to appreciate from the outset is that its discovery shattered our previous notions of animal relationships and of what made animals different, and opened up a whole new way of looking at evolution.

We now know from sequencing the entire DNA of species (their *genomes*) that not only do flies and humans share a large cohort of developmental genes, but that mice and humans have virtually identical sets of about 29,000 genes, and that chimps and humans are nearly 99 percent identical at the DNA level. These facts and figures should be humbling to those who wish to hold humans above the animal world and not an evolved part of it. I wish the view I heard expressed by Lewis Black, the stand-up comedian, was more widely shared. He said he won't even debate evolution's detractors because "we've got the fossils. We win." Well put, Mr. Black, but there is far more to rely on than just fossils.

Indeed, the new facts and insights from embryology and Evo Devo devastate lingering remnants of stale anti-evolution rhetoric about the utility of intermediate forms or the probability of evolving complex structures. We now understand how complexity is constructed from a single cell into a whole animal. And we can see, with an entirely new set of powerful methods, how modifications of development increase complexity and expand diversity. The discovery of the ancient genetic tool kit is irrefutable evidence of the descent and modification of animals, including humans, from a simple common ancestor. Evo Devo can trace the modifications of structures through vast periods of evolutionary time—to *see* how fish fins were modified into limbs in terrestrial vertebrates, how successive rounds of innovation and modification crafted mouthparts, poison claws, swimming and feeding appendages, gills, and wings from a simple tubelike walking leg, and how many kinds of eyes have been constructed beginning with a collection of photosensitive cells. The wealth of new data from Evo Devo paints a vivid picture of how animal forms are made and evolve.

The Tool Kit Paradox and the Origins of Diversity

The stories of shared body-building genes and of the similarities of our genome to that of other animals have slowly been gaining in public awareness. What is generally neglected, however, is how the discovery of this common tool kit and of great similarities among different species' genomes presents an apparent paradox. If the sets of genes are so widely shared, how do differences arise? The resolution of this paradox and its implications are central to my story. The paradox of great genetic similarity among diverse species is resolved by two key ideas that I will develop in the course of the book and will return to repeatedly. These concepts are crucial for understanding how the species-specific instructions for building an animal are encoded in its DNA and how form is generated and evolves. The substance of these ideas has received scant, if any, attention in the general press, but these ideas have profound implications for understanding great episodes in life's history, such as the explosion of animal forms during the Cambrian period, the evolution of diversity within groups such as butterflies or beetles or finches, and our evolution from a common ancestor with chimps and gorillas.

The first idea is that diversity is not so much a matter of the complement of genes in an animal's tool kit, but, in the words of Eric Clapton, "it's in the way that you use it." The development of form depends upon the turning on and off of genes at different times and places in the course of development. Differences in form arise from evolutionary changes in where and when genes are used, especially those genes that affect the number, shape, or size of a structure. We will see that there are many ways to change how genes are used and that this has created tremendous variety in body designs and the patterning of individual structures.

The second idea concerns where in the genome the "smoking guns" for the evolution in form are found. It turns out that it is not where we have been spending most of our time for the past forty years. It has long been understood that genes are made up of long stretches of DNA that are decoded by a universal process to produce proteins, which do the actual work in animal cells and bodies. The genetic code for proteins, a twenty-word vocabulary, has been known for forty years, and it is easy for us to decode DNA sequences into protein sequences. What is much less appreciated is that only a tiny fraction of our DNA, just about 1.5 percent, codes for the roughly 25,000 proteins in our bodies. So what else is there in the vast amount of our DNA? Around 3 percent of it, made up of about 100 million individual bits, is *regulatory*. This DNA determines when, where, and how much of a gene's product is made. I will describe how regulatory DNA is organized into fantastic little devices that integrate information about position in the embryo and the time of development. The output of these devices is ultimately transformed into pieces of anatomy that make up animal forms. This regulatory DNA contains the instructions for building anatomy, and evolutionary changes within this regulatory DNA lead to the diversity of form.

In order to understand the role and significance of regulatory DNA in evolution, I have some ground to cover. One must first appreciate how animals are constructed and the roles of genes in embryonic development. This will form the first half of the book and it holds many rewards in its own right. I will illustrate some general features of animal architecture and trends in the evolution of body design that are shared among different groups of animals (chapter 1). I will describe some of the spectacular mutant forms that led biologists to the tool kit of master genes that regulate development (chapters 2 and 3). We will see these genes in action and how they reflect the logic of and order to the building of animal bodies and complex patterns (chapter 4). And we will learn about the devices in the genome that contain the instructions for building anatomy (chapter 5).

In the second half of the book, I will tie together what we know about fossils, genes, and embryos in the making of animal diversity. I will highlight some of the most important, interesting, or compelling episodes in animal evolution that illustrate how nature has forged many individual designs from a small number of building blocks. I will examine in depth the genetic and developmental foundations of the Cambrian Explosion that produced many of the basic types of animals and body parts we know today (chapters 6 and 7). I will probe into the origins of butterfly wing patterns as splendid examples of how nature invents by teaching very old genes new tricks (chapter 8). I will tell some stories about the evolution of the plumage of island birds and the coat colors on mammals (chapter 9). These are all very satisfying, aesthetically pleasing tales that provide deep insights into the evolutionary process. But they have more direct ramifications, for they are the case studies that reveal the very kinds of processes that shaped human origins. In the final chapters of the book, I will describe the making of our species, most notable for its "beautiful mind" more than any other physical trait (chapter 10). I will trace our beginning from an apelike ancestor 6 million years ago to track the physical and developmental changes that led to *Homo sapiens*. I will discuss the scope and types of genetic changes that have occurred in the course of our evolution and those that are most likely to account for the evolution of traits we most associate with being human.

The Grandeur in a More Modern Synthesis: Act III

The continuing story of evolution may be thought of as a drama in at least three acts. In Act I, almost 150 years ago, Darwin closed the most important book in the history of biology by urging his readers to see the grandeur in his new vision of nature—in how, "from so simple a beginning, endless forms most beautiful have been, and are being, evolved." In Act II, the architects of the Modern Synthesis unified at least three disciplines to forge a grand synthesis. Here in Act III, there is also a special grandeur in the view embryology and evolutionary developmental biology provide into the making of animal form and diversity. Part of it is visual, in that we can now see how the endless forms of different animals actually take shape.

But beauty, in science, is much more than skin-deep. The best science is an integrated product of our emotional and intellectual sides, a synthesis between what is often referred to as our "left" brain (reasoning) and "right" brain (emotional/artistic) hemispheres. The greatest "eurekas" in science combine both sensual aesthetics and conceptual insight. The physicist Victor Weisskopf (also a pianist) noted, "What is beautiful in science is the same thing that is beautiful in Beethoven. There's a fog of events and suddenly you see a connection. It expresses a complex of human concerns that goes deeply to you, that connects things that were always in you that were never put together before."

In short, the best science offers the same kind of experience as the best books or films do. A mystery or drama engages us, and we follow a story toward some revelation that, in the very best examples, makes us see and understand the world more clearly. The scientist's main constraint is the truth. Can the nonfiction world of science inspire and delight us as much as the imagined world of fiction?

One hundred years ago, Rudyard Kipling published his classic *Just So Stories*, a collection of children's tales inspired by his experiences in India. Kipling's enchanting stories ranged from "How the Leopard Got His Spots" and "How the Camel Got His Hump" to "The Butterfly That Stamped," and wove fanciful tales of how some of our favorite and most unusual creatures acquired prominent features. As delightful as the *Just So* explanations are of how spots, stripes, humps, and horns came to be, biology can now tell stories about butterflies, zebras, and leopards that I contend are every bit as enchanting as Kipling's fairy tales. What's more, they offer some simple, elegant truths that deepen our understanding of all animal forms, including ourselves.

Part I

The Making of Animals

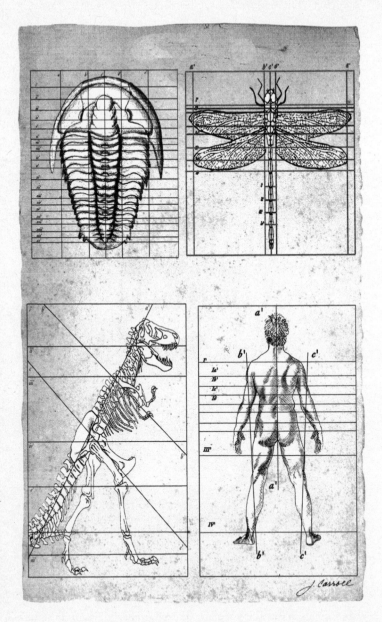

Animal architectures, modern and ancient. JAMIE CARROLL

1

Animal Architecture: Modern Forms, Ancient Designs

It is the mystery and beauty of organic form that sets the problem for us.

—Ross Harrison, embryologist (1913)

THE AMAZING VARIETY of animal forms does not end with those on land or in the sea. Belowground, buried in as little as a few inches of sand or up to several thousand feet of rock, is the story of 600 million years of animal history—from the enigmatic forms of early animals in Canadian shale, to the enormous bodies of dinosaurs in the buttes and valleys of the American West, and the teeth and skull fragments of our bipedal ancestors in the Rift Valley of east Africa. And some of what lies below the ground can be quite surprising given what breathes just above.

I learned this firsthand only recently in, of all places, Florida, a favorite destination for vacationers and retirees seeking sun, entertainment, and relaxation. It is a land of palm trees, soft sandy beaches, graceful pelicans and ospreys, gentle manatees and dolphins, and *Homo sapiens* in plaid pants . . . but also of six-foot-long armadillos,

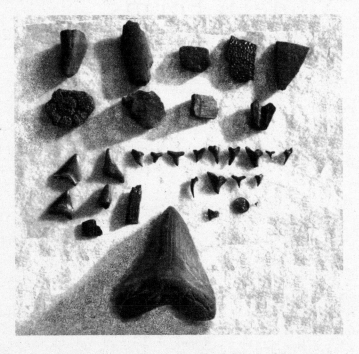

FIG. 1.1 **Fossils from a Florida riverbed.** Mammal bone, turtle shell fragments, and shark teeth abound. Note the variety of shapes and sizes. The largest tooth is from the enormous shark *Charcharadon megalodon.* FOSSILS COLLECTED AND PHOTOGRAPHED BY PATRICK CARROLL

tusked mastodons, sixty-foot-long sharks, camels, rhinoceroses, jaguars, and saber-toothed cats?

Yes, indeed. Well, it depends on where you look.

Journey inland to a river cutting through the sandy soil and a shovel of gravel from the riverbed might contain teeth from any of ten species of shark, from the intricately serrated and curved snaggletooth, to the absolutely terrifying six-inch flesh rippers of the long extinct behemoth *Charcharadon megalodon* (figure 1.1). In the same gravel there will also be bones of Florida's geologically recent past—of tapirs, sloths, camels, horses, glyptodonts, mastodons, dugongs, and other species now vanished.

The diversity of living and fossil animal forms in just this one locale frames the two central mysteries at hand: How are individual forms made? And, how have so many different forms evolved?

At first, the variety of animal forms may seem overwhelming. But there are some general, long established trends in animal design that we can make sense of. In this chapter, we'll search for some generalities in animal architecture and evolution that will help us reduce this mind-boggling diversity to some basic themes.

The Construction of Animals from Building Blocks

A basic theme of animal design becomes obvious when one tries to figure out just what bone or tooth one has found in that shovelful of Florida river gravel. The challenge of the game is both to match the fossil to a species, and also to determine where in the animal it belonged. Why is this so hard? This is one demonstration of a basic fact of animal design. Related animals, such as vertebrates, are made up of very similar parts.

Now say, with a bit of expert help, one is able to figure out that a piece of bone is from a dugong (an extinct sea cow). But if it is a rib, which rib?

Or, if a toe bone is from an extinct horse, which toe is it? From individual bones, it is really difficult to tell. Why this is the case punctuates a second basic fact of animal design—that individual animals are made up of numbers of the same kinds of parts, like building blocks.

FIG. 1.2 **The modular architecture of vertebrates.** Top, a Jurassic salamander about 10 centimeters in length. Bottom, *Camarasaurus*, a Jurassic sauropod dinosaur, almost 19 feet in length. SALAMANDER COURTESY OF NEIL SHUBIN, UNIVERSITY OF CHICAGO; *CAMARASAURUS* COURTESY OF CARNEGIE MUSEUM OF NATURAL HISTORY, ALL RIGHTS RESERVED

Some of these parts can be small, such as an individual toe bone, others gigantic, like the backbones (vertebrae) of some vertebrates. These basic elements are ancient and their proportions maintained over vast differences in body size. Both enormous sauropod dinosaurs and small, delicate salamanders from the Jurassic age (over 150 million years ago) display the same repeating modular architecture of the vertebrate body plan (figure 1.2).

The theme of modular design is by no means limited to vertebrates. The famous fossils of the Burgess Shale, some of the first large, complex animals that populated the Cambrian seas more than 500 million years ago, display all sorts of variations on modular body plans (figure 1.3), as do their living descendants today.

The attraction to these fossils is manifold. Certainly, there is a sense of awe and wonder in seeing and touching extinct beasts that lived in worlds that have long since vanished. But we are also drawn to their *form*. The fossils demonstrate evolution's pervasive use of repeating parts and modular architecture in forging animal designs.

Individual body parts also reflect this theme of modular design. Our limbs, for example, are of similar modular design, each limb built of several pieces (thigh, calf, ankle; upper arm, forearm, wrist) and the hands and feet bear five similar digits (figure 1.4). The modular architecture of the limbs of four-legged vertebrates is also an ancient design, plainly evident in the Jurassic fossils.

Sometimes, the modular design of a structure may not at first be apparent. The complex patterns on a butterfly wing may appear chaotic, but on closer inspection one can see that the overall pattern is also built of repeating motifs. The underside of the blue *Morpho* butterfly has repeating patterns of lines, chevrons, and spots where each of the individual elements are separated by the wing veins (figure 1.5). This shows that each division of the wing enclosed by veins is a unit. The overall pattern is a product of repeating these modular units, each with some variations on the size or shape of the lines, chevrons, or spots within.

The repeating patterns on body structures extend down to very fine

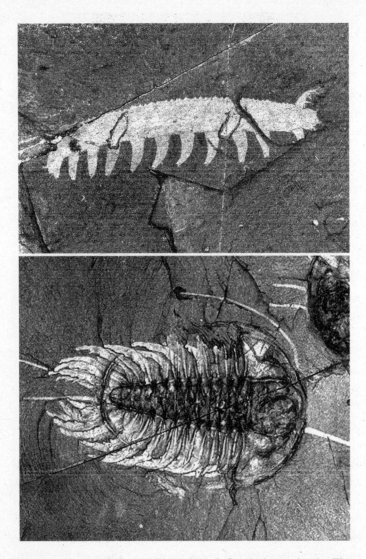

FIG. 1.3 **The modular architecture of Cambrian animals.** The lobopodian *Ayshaeia pedunculata* (above) and trilobite *Olenoides serratus* (below) display repetitively organized, modular body forms. PHOTOS COURTESY OF CHIP CLARK, SMITHSONIAN INSTITUTION

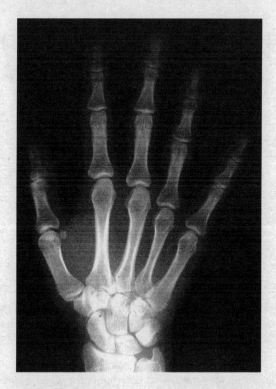

FIG. 1.4 **The modular design of a human hand.** The finger bones revealed by an X ray display a serially reiterated architecture. COURTESY OF JAMIE CARROLL

details, some almost out of range of the naked eye. Such beautiful butterfly wing patterns are actually built from tiny scales on the wing. Each scale is a projection of single cells that are assembled in many rows in the wing. Each scale has its own specific color, like the brushstrokes on a pointillist painting that when combined in a field of thousands and millions of scales, create the overall pattern we admire. The body patterns of fish, snakes, and lizards are also composed of scales (different from those of butterflies) arrayed in a regular geometric pattern. The reflective or refractive properties of scales depend on even

FIG. 1.5 **Serially repeating designs within butterfly wings** are shown in the underside of this *Morpho* butterfly. Each wing is made up of serially reiterated subunits demarcated by two veins and the wing edge. Each subunit contains variations of the same elements—eyespots, bands, and chevrons. BUTTERFLY GIFT OF NIPAM PATEL, PHOTO BY JAMIE CARROLL

finer cellular microanatomy that determines the wavelengths of light that are absorbed or reflected (figure 1.6).

From just these few descriptions, we can begin to appreciate the immense task of development—to build large, complex animals beginning with only a tiny single cell. There are millions of details, and the details count. A small shift in an early process would have a cascade of later effects. What processes can assemble both a massive dinosaur and paint the delicate details of a spot on a butterfly?

Given such enormous differences in scale, and such great variety in animal forms, it would seem that the details of development would present what molecular biologist Gunther Stent described only twenty years ago as "a near infinitude of particulars which have to be sorted out case by case." But biologists have been surprised and delighted to find there are generalities we can make about form and, fortunately,

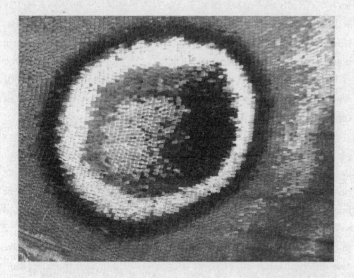

FIG. 1.6 **Repetition on a fine scale.** The scales of a butterfly
wing are like the strokes of a pointillist painting, each stroke being
a single scale of a particular color; collectively they form geometric
pattern elements. PHOTO BY STEVE PADDOCK

these generalities also extend far more deeply than outward appearances, into the genetic machinery of development. I'll start with the
outward similarities here and work my way down to the genes that do
the job in the next two chapters.

Evolution as Variations in Number and Kind

The modular and repetitive aspects of animal design reflect an order to
animal forms. Anatomists have long appreciated that no matter how
diverse their outward appearance, animal bodies and their parts are
constructed along some perceivable themes. More than a century ago,
some of these themes were formally defined by the English biologist

William Bateson. His perspective turns out to be a very helpful framework for thinking about the logic of animal design and understanding how variations on basic themes evolve.

Bateson recognized that many large animals were constructed of repeated parts, and many body parts themselves were constructed of repeated units. In considering particular groups of animals, it appeared that some of the most obvious difference between members of a group were in the *number* and *kind* of repeated structures. For example, while all vertebrates have a modularly constructed backbone made up of individual vertebrae, different vertebrates possess different numbers and kinds of vertebrae. The number of vertebrae from head to tail differs greatly, from fewer than a dozen in frogs, to thirty-three in humans, and up to a few hundred in a snake (figure 1.7). There are also different kinds of vertebrae such as cervical (neck), thoracic, lumbar, sacral, and caudal (tail) vertebrae. The main differences between these types in any one animal are their size and shape and the presence or absence of structures attached to them, such as ribs. There is great diversity in the number of each type in different vertebrates.

A similar pattern applies to arthropod form and diversity. Arthropod bodies are made up of repeating segments, which in the trunk (behind the head) may vary from about eleven segments in insects to dozens in centipedes and millipedes. Groups of segments are distinguished from one another (e.g., the thoracic and abdominal segments) by their size and shape and especially by the appendages that project from them (e.g., the thoracic segments in insects each bear a pair of legs while the abdominal segments do not).

These two groups of animals have successfully exploited every environment on earth (water, land, and air) and are the most complex animals in terms of anatomy and behavior. Both groups are constructed of repeated assemblages of similar parts. Is there a connection between modularity of design and the success in evolutionary diversification? I certainly think so. The challenge for biologists has been to figure out

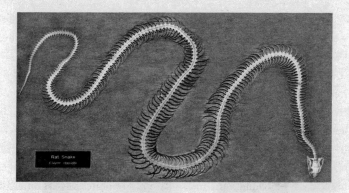

FIG. 1.7 **Snake skeleton.** Hundreds of repeating vertebrae and ribs make up the snake body form. COURTESY OF DR. KURT SLADKY, UNIVERSITY OF WISCONSIN

how these animals are built, beginning from just a single cell, and how all sorts of variations on a body design evolve. The modular construction of vertebrates and arthropods, and their variation in the number and kinds of modules, are important clues to the processes involved.

Body parts that are often modular and constructed of similar units often differ between species largely in number and kind. The limbs of four-legged vertebrates (tetrapods) usually bear one to five digits. We recognize five distinct types of digits on our own hands (thumb, forefinger, etc.) and feet. The similarities among digits are obvious, the differences largely a matter of size and shape. The tetrapod limb has been adapted to many functions in a great variety of designs, and the basic five-digit design has persisted for more than 350 million years, although digit number has evolved extensively such that anywhere from one to five digits may be present (for example, camels have two toes, rhinos have three, etc.). The variations on the tetrapod theme are spectacular, as a sample of X rays among vertebrates highlights (figure 1.8). Interestingly, closely related animals can differ widely; some groups have evolved many species that differ from one another in digit number.

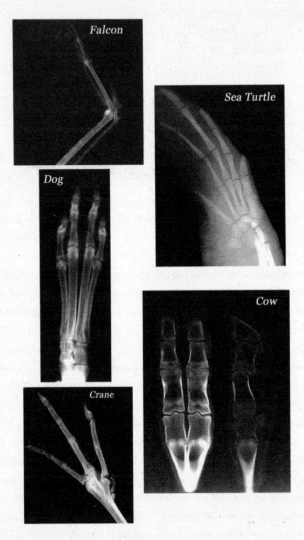

FIG. 1.8 **Diversity of vertebrate limb forms.** All vertebrate limbs are variations on a common design in which the number, size, and shape of elements (such as digits) differ. COURTESY OF DR. KURT SLADKY, UNIVERSITY OF WISCONSIN; SEA TURTLE COURTESY OF DR. CRAIG HARMS, NORTH CAROLINA STATE UNIVERSITY

Homology, Serial Homology, and Williston's Law

When comparing body parts between species, it is important to know whether one is comparing the same body part that might have changed in different ways, or one is comparing parts in a series, where the one-to-one relationship may be obscured. For example, the forelimbs of salamanders, sauropods, mice, and our arms are all *homologs*. This means they are the same structure modified in different ways in each species. They are derived from a common ancestral forelimb. Hindlimbs, our legs and the hind legs of four-legged vertebrates, are also homologs. With respect to each other, forelimbs and hindlimbs are *serial homologs*, structures that arose as a repeated series and have become differentiated to varying degrees in different animals. Vertebrae and their associated structures (ribs); tetrapod forelimbs and hindlimbs; digits; teeth; the mouthparts, antennae, and walking legs of arthropods; and the fore- and hindwings of insects are also serial homologs.

Changes in the number and kind of serial homologs are a principal theme in animal evolution. Let me drill this home with a couple of more examples of familiar structures. If you are a seafood lover, chances are you have dissected a lobster. While dismembering it, perhaps you might have noted the modular design and admired the great variety of body appendages (figure 1.9). There are several aspects to lobster construction that reflect the general themes of modularity and serial homology. First, the body is organized into a head (with the eyes and mouthparts), a thorax (with walking legs), and a long tail (yum!). Second, different sections of the body possess numbers of specific appendages (antennae, claws, walking legs, swimmerets). And third, each jointed appendage is itself segmented, and different kinds of appendages have different numbers of segments overall (compare a claw with a walking leg). If you were feeling adventuresome and dissected an insect or a crab, you'd see some general similarities in body

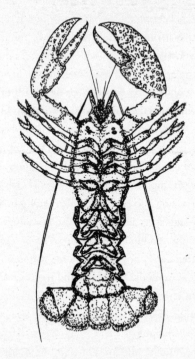

FIG. 1.9 **The diversity of the serially repeated appendages of a lobster.** The antennae, claws, walking legs, swimmerets, and tail structures are all modifications of a common limb design. DRAWING BY JAMIE CARROLL

organization, segmentation, and appendages but, again, differences in the number and kind of serially homologous structures.

A second example of serially homologous parts would be the teeth you used to mince and crush that lobster. Our jaws host a variety of teeth (canines, premolars, incisors, molars, etc.). Again, one of the obvious differences among all sorts of vertebrates are the number and kind of teeth. Primitive reptiles, like great marine forms, had a mouth full of mostly similar teeth, but later species evolved different kinds of teeth, adapted for biting, tearing, and compacting food. The differences in dental hardware reflect differences in diet, with carnivores bearing incisors and canines and grazers bearing mostly molars (figure 1.10). We differ from our primate relatives in our dentition (figure 1.11). You may be aware that teeth make hardy fossils and such finds have played a major role in deciphering the identity and lifestyle of our ancestors.

The evolutionary trends in the number and kinds of repeated structures are so pronounced that the paleontologist Samuel Williston

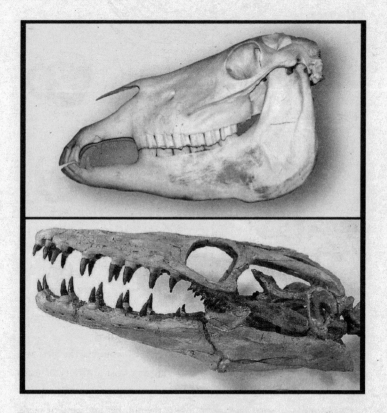

FIG. 1.10 **Teeth in a primitive vertebrate**. In mosasaurs (bottom), all teeth appear mostly similar, whereas later vertebrates (top; here a horse) had teeth of distinct types. RECONSTRUCTION OF *PLATECARPUS PLAIFRONS* COURTESY OF MIKE EVERHART, OCEANS OF KANSAS PALEONTOLOGY

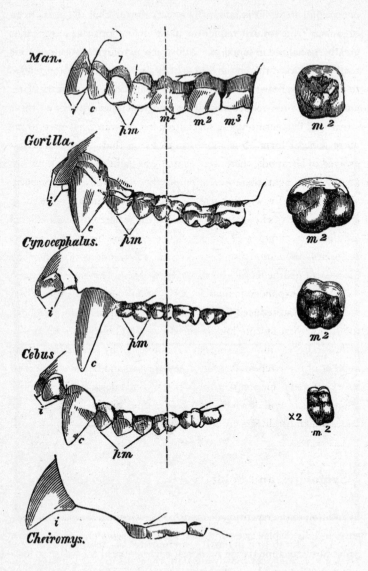

FIG. 1.11 **The diversity of primate dental hardware.** Primates differ in the number and shape of canine, premolar, and molar teeth. FROM T. H. HUXLEY, *MAN'S PLACE IN NATURE* (1863)

declared in 1914, "it is [also] a law in evolution that the parts in an organism tend toward reduction in number, with the fewer parts greatly specialized in function." Williston was studying ancient marine reptiles. He noted that in the course of evolution, earlier groups tended to have large numbers of similar serially reiterated parts, but that later groups exhibited reduced numbers and specialized forms of these structures. Furthermore, the specialized pattern rarely reverted to the more general form. One interesting case is that when digits first evolved in tetrapods, there were as many as eight digits per foot. But among these eight, there were no more than five types, which eventually reduced to five digits that were specialized, or further reduced, in later species. Laws in biology are few, and those dared to be articulated are almost certain to be broken by some organisms. Yet Williston's Law is a useful observation that seems to pertain to trends in more than just the ancient marine reptiles he was writing about. The trend appears to be that once expanded in number, serial homologs became specialized in function and reduced in number. The specialization of vertebral, tooth, and digit morphology in vertebrates, and of legs and wings in arthropods, was in fact generally accompanied by a reduction in the number of these repeated structures. Williston and Bateson appear to have captured some simple truths about animal design and evolution, allowing us to boil down the vast history and variety of some of the largest and most diverse groups into some generalities.

Symmetry and Polarity

In addition to the repetition of modular parts, animal bodies and body parts usually display two additional features—*symmetry* and *polarity*. Most familiar animals are bilaterally symmetrical in that they have matching right and left sides with a central axis of symmetry running down the middle of the long axis of the body. This design also imposes a front/rear orientation to animals and has enabled the evolution of many efficient modes of locomotion. Some animals exhibit other symmetries, such as the pentaradial (five-fold) echinoderms, a group including sea urchins, sand dollars, and a spectacular variety of other species (figure 1.12). The axes of symmetry in an animal are clues to how the animal is built.

So, too, is the polarity of an animal and its parts. In most animals there are three axes of polarity: head to tail, top to bottom (back and front in ourselves, since we stand up), and near to far from the body (in reference to structures that project from the main body—such as a limb whose parts are organized perpendicular to the main body). Individual structures also have polarity. Think of the hand, which has three axes oriented by the thumb to pinkie, back to palm, and wrist to fingertip directions.

How Is Form Encoded in the Genome?

Modularity, symmetry, and polarity are nearly universal features of animal design, certainly of larger, more complex animals such as butterflies and zebras. These features and the evolutionary trends noted by

FIG. 1.12 **Other symmetrical animal forms.** Echinoderms such as sea urchins (left), sand dollars (center), and starfish (right) are radially symmetrical. DRAWING BY JAMIE CARROLL

Williston and Bateson suggest that there is order and logic to animal architectures. They suggest that underneath the great variety of animal forms, there are some general "rules" to be discovered about how animals are built and evolve.

In the course of this book, I will focus on four main questions:

1. What are some of the major "rules" for generating animal form?
2. How is the species-specific information for building a particular animal encoded?
3. How does diversity evolve?
4. What explains large-scale trends in evolution, such as the change in number and function of repeated parts?

Where do we look for these rules and instructions? In DNA. In the entire complement of DNA of a species (the genome), there exists the information for building that animal. The instructions for making five fingers, or two eyespots, or six legs, or black and white stripes are somehow encoded in the genomes of the species that bear these traits. Does this mean there are genes for fingers, genes for spots, genes for stripes, etc.? I will focus on how anatomy is encoded in the genome in the first part of the book.

I'll tackle evolutionary diversity in the second part of the book. Somehow, different species with three versus four fingers, two versus seven eyespots, six versus eight legs, or that are all black or all white, must have different instructions encoded in their DNA. Evolution of form is ultimately then a question of genetics. But in order to understand how genes sculpt all of this breathtaking animal beauty, we will first look to monsters for some critical clues.

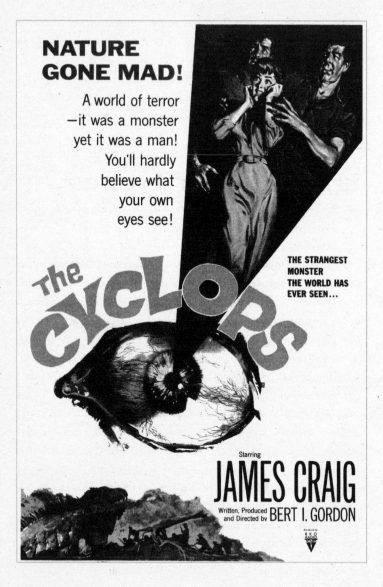

Poster for the movie *The Cyclops* (1956). B & H PRODUCTIONS, INC.

2

Monsters, Mutants, and Master Genes

"Do you know, I always thought Unicorns were fabulous monsters, too? I never saw one alive before!"

"Well, now that we *have* seen each other," said the Unicorn, "if you'll believe in me, I'll believe in you. Isn't that a bargain?"

—Lewis Carroll,
Through the Looking Glass (1872)

CREATURE FEATURE was a staple of Saturday afternoon television when I was a kid. My best friend, Dave, was addicted to the show. He'd hole up in his basement with the curtains drawn, lights off, a baseball bat at his side, and all sorts of contraptions rigged to the door and windows in case one of the featured monsters happened to visit him in the middle of the show. He'd watch hours of Godzilla, Dracula, the Mummy, or worse. Dave would later recount the plot lines for us and speculate on the relative strengths and unique powers of all the beasts. Fueled by an active imagination, which was probably also stoked by the five-gallon tin of popcorn and the can of Betty Crocker frosting that were his standard refreshments, these creatures had become almost *real* to him.

Our fascination with and terror of monsters are universal and ancient. From Greek mythology to B movies, writers have imagined all sorts of giants, hybrids, and ghoulish creatures. I didn't share Dave's appetite for monster movies (or cans of frosting, for that matter), but monsters have played an important role in advances in embryology. One of the most successful approaches to figuring out how animal forms develop correctly has been the study of dramatic monsters with the wrong number of parts, or parts in the wrong places. Some of these forms were man-made creations, some the product of accidents and injuries during gestation, and others the result of rare mutational events in nature. The insights gained from study of these kinds of monsters have recently converged to reveal specific mechanisms underlying the assembly of all animal bodies and body parts.

The Myth and Reality of Cyclops

I never bought into tales of the dead coming alive, humans transformed into bats or flies, gigantic skyscraper-scaling gorillas, creatures that were half human, half horse/goat/snake/fish/whatever, fire breathers, or invisible bodies. I put them all into the same category of dark fairy tales. I did the same with monsters with one central eye, but here is where I might have been too hasty to dismiss a creature.

While I was vaguely familiar with the mythology of Cyclops, I did not know that animals with one central eye were actually well-known to science. In fact, at one time in Utah, 5 to 7 percent of newborn sheep

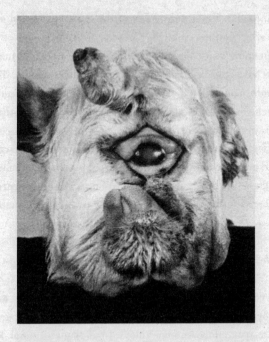

FIG. 2.1 **Cyclopic lamb.** Caused by exposure of mother at critical period to cyclopamine, a toxin produced by the plant *Veratrum californicum*. PHOTO COURTESY OF DR. LYNNE JAMES, POISONOUS PLANT RESEARCH CENTER, LOGAN, UTAH

were afflicted with cyclopia, a lethal malformation in which they bear a single central eye, lack most nasal and jaw structures, and have incompletely developed brain hemispheres (figure 2.1). The formal term is *holoprosencephaly*, meaning single forebrain, and the key defects are that the forebrain and eyes fail to become separated into two symmetrical structures.

The high frequency of cyclopia in sheep was eventually found to be associated with the presence of a plant, the lily *Veratrum californicum*, in the pastures in which their mothers grazed. Ingestion of this plant during a period of gestation (around the fourteenth day) was the critical factor. It turns out that the plant produces a chemical called cyclopamine that has a teratogenic (from the Greek *teras*, meaning monster) effect on the developing embryo.

Cyclopamine is just one of many known teratogens. Many other chemicals have adverse affects on development of embryos. The drug thalidomide, originally developed to treat nausea during pregnancy, is probably the most notorious, having caused thousands of birth defects in humans in the late 1950s and early 1960s. While these molecules have been known for decades, there was no progress in understanding how they acted until the recent convergence of embryology with molecular biology. These advances have stemmed from more specific experiments, especially through the manipulation of embryos and genes.

Newt Lips and Chicken Wings

Over the past century, biologists have used scalpels, needles, tweezers, and all sorts of tools to chop, strangle, burn, blend, spin, and prick embryos to try to uncover some rules for building an animal. Pioneers in embryology relied entirely on physical methods to move and remove cells and then observe what unfolded in the embryo. From these crude tortures, some dramatic monsters were produced whose striking features revealed a few central principles governing the organization of developing animals.

Foremost among these pioneers was Hans Spemann, the first, and for a span of more than sixty years, the only embryologist to win the Nobel Prize (but as we will see embryologists have lately been catching up). One of the very first illuminating experiments he carried out was to test whether the first two cells of a newt embryo had similar or different properties. Spemann used a fine baby's hair, taken from his own daughter, to tie off and separate individual halves of the embryo. The cells on each side of the knot gave rise to normal newt tadpoles, demonstrating that the two halves of early amphibian embryos could give rise to two entirely identical animals.

When Spemann divided the egg differently by tying it perpendicular to the furrow between the two cells of the embryo, he obtained a dramatically different result. Only one side made a normal tadpole, while the other made a disorganized mess of belly tissue. This eventually led to the recognition that a region of the embryo, called the dorsal lip of the blastopore, was critical for the organization of the embryo. If this region of the embryo was removed, the embryo formed a blob of tissue lacking structures that normally form on the top (dorsal) side of the animal. More spectacularly, if this dorsal lip region was transplanted to the presumptive belly region of another developing embryo, it organized a second embryonic axis and *two* embryos formed that were joined together (figure 2.2)! Spemann dubbed this region "the organizer" because he deduced that it organized the dorsal part of the embryo into neural structures and could initiate the development of another embryonic axis.

The spectacular effects of the Spemann organizer revealed that one way order is brought about in development is by interactions between one part of an embryo and other parts. Some other organizers with dramatic properties have been discovered that show that this principle works on many scales in development: across the whole embryo, within individual body parts, and right down to intricate details of patterns. Let's look at two more organizers that illustrate these dramatic activities.

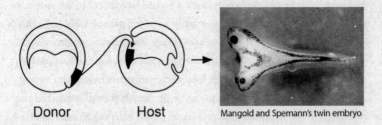

Donor   Host   Mangold and Spemann's twin embryo

FIG. 2.2 **Induction of a second axis and embryo in a frog tadpole.** Transplantation of "organizer" tissue to a different site induces the formation of a conjoined embryo. PHOTO COURTESY OF HIROKI KURODA AND EDDY DE ROBERTIS, UCLA

The formation of limbs has long held the fascination of embryologists. Beginning from just a small bud on the flank of the embryo early in development, the limb takes shape in many stages. In a three-day-old chicken embryo, the bud is initially only about 1 millimeter long and 1 millimeter wide but this will grow more than a thousand-fold by the time the chick hatches. In the intervening period, this tiny pad of tissue will grow outward, lengthen, and develop bone, cartilage, muscles, tendons, digits, and feathers in a beautiful display of the coordinated processes of development. Perhaps most striking is the orderly formation of cartilage elements (which will later be replaced with bone). The cartilage is formed around condensations of cells and laid down in order from the shoulder to the wrist and then to the digits. The whole process can be seen with special stains (figure 2.3). The order of events in limb development and the polarity of the digits suggests that, like the embryo, there must be some sort of system that cues cells as to what they will become.

Several decades ago, John Saunders, another pioneer embryologist, discovered an organizer of polarity in the chicken embryo wing bud. A chicken wing normally has three digits, which we can identify due to their size and shape as digits 2, 3, and 4 (in order from the front to the back of the wing; digits 1 and 5 do not form in the wing). When Saunders transplanted a chunk of tissue from the posterior part of the growing limb bud (near where digit 4 would originate) to an anterior position (where digit 2 would normally appear), a wing with extra digits formed. The digits were a mirror image duplication of the normal pattern—that is, instead of the 2 3 4 sequence of wing digits, a pattern of 4 3 2 3 4 formed (figure 2.4). The wing's mirror image polarity suggested that cells in the posterior zone of the limb bud organized the polarity of the digit sequence (4, 3, 2), such that when moved elsewhere, the 4, 3, 2, sequence was induced in a new place.

The realms of influence of Spemann's organizer and the zone of polarizing activity (ZPA) in the chick limb are pretty large, affecting the development of the entire embryo or a large body part. But organ-

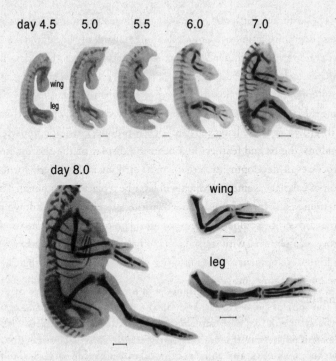

day 4.5   5.0   5.5   6.0   7.0

wing
leg

day 8.0

wing

leg

FIG. 2.3 **Limb formation in a chicken.** The wing and leg buds grow dramatically over several days of embryonic development. The laying down of cartilage, which precedes bone deposition, is visualized with a special stain and proceeds in order from upper limb parts to the digits. Note that the details of wing and leg anatomy differ in detail. PHOTOS COURTESY OF JOSEPH J. LANCMAN AND JOHN FALLON, DEPARTMENT OF ANATOMY, UNIVERSITY OF WISCONSIN

izers have been found that act in finer scales. In 1980, Fred Nijhout of Duke University showed that the eyespot patterns on butterfly wings were also induced by an organizer. When Nijhout killed a tiny patch of cells that would form the center of the eyespot, no eyespot formed. More interesting, he found that when this small group of cells was isolated from the developing butterfly wing in the first day of the chrysalis stage and transplanted to a site elsewhere on the wing, a new eyespot

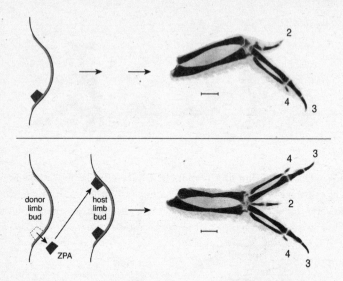

Fig. 2.4 **Induction of polydactyly in a chicken.** Transplantation of the zone of polarizing activity in the developing wing bud from a posterior site to a new anterior position induces extra digits with opposite polarity to the normal digit pattern. PHOTOS COURTESY OF JOSEPH J. LANCMAN AND JOHN FALLON, DEPARTMENT OF ANATOMY, UNIVERSITY OF WISCONSIN

now appeared (figure 2.5). Only cells at the future center of the eyespot had this property. Nijhout dubbed the eyespot organizer the "focus."

All organizers share the property of influencing the formation of pattern, or *morphogenesis*, in tissues or cells. The basic interpretation of their special activity is that the cells of organizers produce substances that can influence the development of other cells. Such substances have been dubbed morphogens. The effects of organizers depend upon their distance from target cells: cells nearby are most affected whereas cells farther away in the newt embryo, limb bud, or butterfly wing are not (or less) affected. It has long been thought that

FIG. 2.5 **Induction of eyespots in a butterfly.** Transplantation of cells at the center of a developing eyespot to other locations in the developing wing induces an eyespot in those locations. PHOTO COURTESY OF DR. H. FREDERIK NIJHOUT, FROM HIS *THE DEVELOPMENT AND EVOLUTION OF BUTTERFLY WING PATTERNS*, USED BY PERMISSION OF SMITHSONIAN INSTITUTION PRESS

morphogens produced in one site diffuse outward and form *concentration gradients* from their source. The idea then is that cells surrounding the source respond according to the amount of morphogen they experience. For example, in the chicken wing bud, cells close to the ZPA develop the posterior type of digit (digit 4), and those farther away progressively develop more anterior type of digits (digits 3 and 2 respectively). In the butterfly eyespot, the concentric rings of differently colored scales are thought to reflect different responses to different levels of the focal morphogen at different distances from the focal source.

The morphogens responsible for the activity of organizers were some of the most sought "Holy Grails" of embryology. The major difficulty that retarded further advances was that the organizer activity was a property of collections of cells. Any cell makes thousands of substances and it was always possible that more than one substance was responsible for organizer activity. While transplantation was a powerful tool, embryologists needed some way to find morphogens in the complex soup of cell biochemistry. They would wait decades.

Hopeful Monsters

The animals created by Spemann, Saunders, and Nijhout were man-made monsters with duplicated axes or extra digits or altered wing spots. But these sorts of abnormalities were not unknown in nature. In fact, Bateson, in his 1894 treatise *Material for the Study of Variation,* catalogued and described a whole parade of "monsters" from across the animal kingdom that displayed extra, missing, or altered parts. Bateson culled from museums, collectors, and anatomy departments across Europe a menagerie of oddities including: a sawfly and a bumblebee with legs in place of their left antennae; crayfish with extra oviducts; butterflies with missing or extra eyespots; frogs with extra vertebrae or transformations of vertebral types; and much more (figure 2.6).

Bateson divided these abnormalities into two basic types: those in which the number of repeated parts was altered and those in which one body part was transformed into the likeness of another. He called the lat-ter variants *homeotic* (from the Greek *homeos* meaning same or similar), and this will be a very important term to remember. Bateson's motivation for collecting these oddities was to show that leaps in morphology can occur in nature and thus could be the basis of evolutionary change. I have to say right off the bat that, as intuitive and appealing as Bateson's rea-soning may first appear, biologists generally believe with very good rea-son that the notion of evolution making such large leaps in a single bound is very, very unlikely. The fact that such variations arise does not mean that these are founders of new types or species. Rather, from what we now know these monsters are almost certainly misfits that will be swept away by the power of natural selection with no chance of passing on their traits. This notion of "hopeful monsters" giving rise to new forms in one single bound has been very difficult to dispel, particularly in the popular scientific press (the BBC even produced a program with this title a few years ago, despite my pleas with the producer that it was a discredited idea). It is a seductive notion, but without any merit. In the course of this book we shall see that there is no support for, nor any need to invoke, hopeful monsters as agents of evolution.

Perhaps the most obvious limitation presented by Bateson's catalog is that most examples presented defects in only one of a pair of structures. While provocative, these one-of-a-kind museum pieces were rare finds

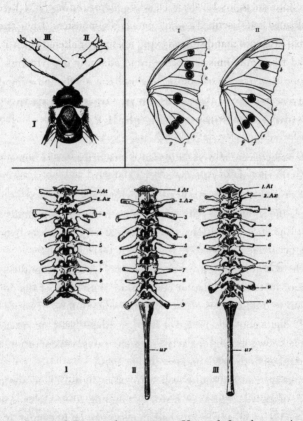

Fig. 2.6 **Some of Bateson's monsters.** Upper left, a homeotic transformation in a sawfly where one antenna is transformed into a leg. Upper right, eyespots missing from a butterfly wing. Lower panels, variation in vertebrae and their projections in a frog. FROM W. BATESON, *MATERIALS FOR THE STUDY OF VARIATION* (1894)

and their causes were unknown. It was important to understand, for example, whether such forms were heritable or might be the result of physical damage to the embryo while it was forming (and thus would not be inherited). It did turn out that Bateson's kinds of monsters were informative, not so much for telling us about the true cause of evolution, but for insights into development that bear on evolution. As foretold in one of my favorite essays by the late Stephen Jay Gould, one that in fact influenced a change in direction early in my scientific training, Bateson's monsters were "helpful" scientifically, but hopeless as individuals.

How Many Fingers? Human Digit Variation from Anne Boleyn to Baseball Pitchers

Bateson's collection of variations also included cases of humans bearing extra ribs, men with one or a pair of extra nipples, a spectacular case of eight fingers arranged in mirror image symmetry on a left hand, and individuals with extra digits on one or both hands (figure 2.7). The latter conditions, termed polydactyly, are actually not that rare, occurring about 5 to 17 times per 10,000 live births.

There is quite a range of degrees of polydactyly, from the appearance of just an extra flap or bud of skin on the side of the pinkie or thumb to duplications of the nail, individual bones, or entire digits. Extra digits may be separate or fused to other digits; the latter condition is known as synpolydactyly. In some cases, the condition is fully bilateral on both hands and feet (figure 2.8).

Humans can fare quite well with extra digits. There are famous cases of polydactylism in history, including Anne Boleyn, wife of Henry VIII, who apparently had an extra nail on one hand. It is also reported that King Charles VIII of France and Winston Churchill may have had extra digits. Antonio Alfonseca, a relief pitcher for the 2003 baseball World Series Champion Florida Marlins, has six digits on both hands and feet. The extra finger does not affect his pitching grip,

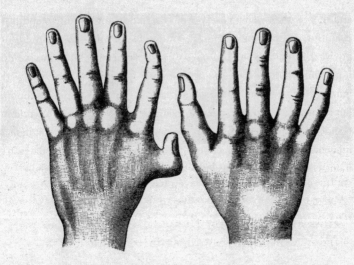

FIG. 2.7 **Polydactyly of a human hand.** FROM W. BATESON, *MATERIALS FOR THE STUDY OF VARIATION* (1894)

so it does not affect his success on the mound. It does appear, however, to offer some psychological advantage as opponents often refer to batting against Alfonseca as "facing the six-fingered man."

Polydactyly is often heritable and pedigrees of polydactylous families are well-known. Indeed, it is reported that in a region of Turkey near Ephesus called Altiparmak, some families have taken the last name Altiparmak, meaning six-fingered ones.

Polydactyly is known widely throughout vertebrates, especially in cats, mice, and chickens. It is striking that similar digit patterns can occur in different animals, including humans, and they can be induced by experimental manipulations or inherited. This suggested that there could be some mechanisms in common for generating extra human fingers and chicken digits. But no progress was made into the mechanisms underlying digit number and pattern until advances were made in understanding some spectacular mutants in animals that have no fingers or toes—the humble fruit fly.

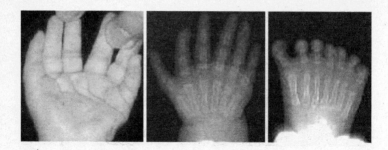

FIG. 2.8 **Polydactyly on both hands and feet.** This patient has six fingers on each hand and seven toes on each foot. PHOTOS COURTESY OF DR. ROBERT HILL, MRC HUMAN GENETICS UNIT, EDINBURGH, SCOTLAND; FROM *PROCEEDINGS OF THE NATIONAL ACADEMY OF SCIENCE,* USA 99 (2002): 7548

Frankenflies

In order to make further progress into what monsters could teach us about the rules of development, a continual supply of abnormal types was needed, monsters that would breed true in the laboratory such that their offspring and subsequent generations would exhibit the same characteristics. In 1915, geneticist Calvin Bridges obtained the first true breeding homeotic mutant in the fruit fly *Drosophila melanogaster*, which was then just beginning to become a leading species for genetic investigations. Bridges isolated a spontaneous mutation that caused the tiny hindwings of the fruit fly to resemble the large forewings. He dubbed this mutant *bithorax*. Subsequently, several more homeotic mutants were identified in *Drosophila*. For example, a rather spectacular mutant *Antennapedia* causes the development of legs in place of the antennae on the head (figure 2.9).

It is remarkable how these homeotic mutants can so completely transform one structure into another. It is not that development is stunted or fails, but that the fate of an entire structure is altered, such that a part is put in the wrong place or the wrong number of parts

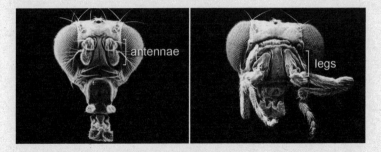

FIG. 2.9 **Homeotic mutant fruit fly.** Left, a normal fly head with antennae; right, an *Antennapedia* mutant fly in which the antennae are transformed into legs. PHOTOS COURTESY OF DR. RUDY TURNER, INDIANA UNIVERSITY

form. Crucially, the transformation is of one serial homolog into the likeness of another (antenna to leg, hindwing to forewing). They are also so intriguing because each transformation is due to a mutation at a *single* gene. In *Drosophila*, only a small number of "homeotic" genes give homeotic forms when they are mutated, indicating that a small number of "master" genes govern the differentiation of serially homologous body parts in the fly.

The spectacular effects of homeotic mutants inspired what would become a revolution in embryology, and then another in evolutionary biology. But in order to appreciate their meaning and the insights they have to offer, we have to dig deeper to understand how these master genes work. How can one gene affect one whole structure and not another? What do the genes encode that have such large effects on animal bodies? Perhaps your first response is: "A fruit fly? Why should I get excited about a fruit fly?" The answers to all of these questions unfold from understanding more about DNA and how genes work, and along the way we'll learn some surprising discoveries about the makeup of different animal genomes.

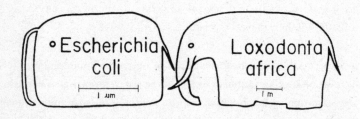

Whimsical representation of Monod's famous quip. COURTESY OF
DR. SIMON SILVER, UNIVERSITY OF ILLINOIS–CHICAGO

3

From *E. coli* to Elephants

What is true for *E. coli* is also true for the elephant.

—Jacques Monod, Nobel laureate

"WE WISH TO SUGGEST a structure for the salt of deoxyribose nucleic acid (DNA). This structure has novel features which are of considerable biological interest." So began James Watson and Francis Crick's paper in 1953 that announced their new and correct model for the genetic material. DNA is a universal, the basis of heredity in all six kingdoms of organisms—bacteria, archeabacteria, protists, fungi, plants, and animals. Within a little more than a decade, another universal, the genetic code, would be solved. But would there be more universals to discover?

The cells, tissues, and organs (where they exist) of members of each kingdom are vastly different in important ways. And there is great diversity within any kingdom; animals for instance span from tiny planktonic species to enormous sea and land mammals. The classification of organisms, the assignment into like and unlike, has largely been driven by consideration of form. So the long-standing assumption has been the greater the disparity in form, the less, if anything, any two species would have in common at the level of their genes.

By the end of this chapter, you will appreciate that looks can be deceiving. This is a chapter with many "ahas." It includes the stories of several powerful—indeed, beautiful—discoveries that have shaped and reshaped our picture of how animals have evolved. The recurring theme is of deep, unexpected connections among different forms. I will begin with some insights from a simple microscopic bacterium that reveal the basics of gene logic. We will ascend from there to more complex beasts and the homeotic genes in fruit flies, and then branch out to the entire animal kingdom.

Proteins, DNA, and Gene Logic

Think about some of the different kinds of cells in our bodies, what they do, and how they do it. Red blood cells carry oxygen to tissues, cells in our digestive organs process our food, neurons carry electrical impulses around our nervous system, and muscles move our body parts. The key to the specialization of these cell types is their selective production of *proteins*, the class of molecules that do all of the work in the body. Red blood cells make huge quantities of the oxygen-binding protein hemoglobin, cells in the pancreas pump out trypsin and other proteins that break down food into usable components, neurons make proteins that set up electrical potentials, and muscle cells make proteins that form into long fibers whose contraction generates force. However, while these cells are devoted to their specialized tasks, they all contain the very same genetic information, the same molecules of DNA. Somehow, each cell type has become different by making some proteins and not others. The selective production of proteins in some places and not others, or at some times and not others, is fundamental to the making of complex organisms.

Before biologists could understand how different cells are made in an animal, they had to solve the basic riddle of how genetic information is stored, copied, and decoded in simple organisms, such as the intestinal bacterium *Escherichia coli*. There are many strains of *E. coli*, some that are good for us and others that are very dangerous. For molecular biologists, *E. coli* has been a marvelous ally that has taught us many of the basic rules about the mechanisms and logic of how genes and proteins work. The insights from this simple bacterium provide a critical foundation for thinking about the development and evolution of more complex creatures.

The mystery that sparked biologists' early interest in *E. coli* was the phenomenon of enzyme induction. *E. coli* loves the simple sugar glucose, but it can break down and use other sugars if glucose is not available. Lactose is a sugar that is broken down into glucose and galactose by an enzyme called beta-galactosidase. When *E. coli* is grown on glucose or other sources of carbon, very little beta-galactosidase is present and the enzyme is made at a slow, almost undetectable drip. *E. coli* doesn't waste its energy making enzymes it doesn't need or can't use. But when lactose is added to a bacterial culture and glucose is absent, the rate of enzyme production is cranked up a thousand-fold, and its presence can be detected in just three minutes. Somehow, the bacterium senses the presence of lactose and is induced to make the right enzyme when it is needed. How can such a simple cell "know" what enzymes to make? How is the right enzyme induced by the appearance of the very compound it breaks down?

The answers to these questions were worked out by François Jacob and Jacques Monod, who along with André Lwoff shared the Nobel Prize in 1965 for their discoveries. No ivory tower theoreticians among these three. Jacob and Monod teamed up at the Pasteur Institute after World War II, during which Monod was a leader in the French Resistance, Lwoff gathered intelligence and occasionally sheltered downed aviators in his apartment, and Jacob was a medic with the Free French through the Africa campaign and was severely wounded in Normandy in August 1944. The historical setting at the outset of their work, the fundamental importance of the implications of their discoveries, and the extraordinary character of these three individuals make the story of enzyme induction and gene logic in bacteria one of the most dramatic in the history of modern biology.

In order to appreciate how enzyme induction works and its implications for more complicated organisms, we have to grasp the basics of the structure and function of DNA, RNA, and proteins. I appreciate that they may be unfamiliar or daunting, but the biological logic emerges when one is able to picture these components in action, and to understand their different roles and interactions. And, most important, there are some very big discoveries to be revealed soon, the impact of which depends upon understanding the different classes of molecules that living organisms are made of.

The relationship between DNA, RNA, and protein is that DNA is a template for the making of RNA, and RNA is in turn the template for the making of proteins. Genetic information stored in DNA is thus decoded in two steps to produce the proteins that do the actual work in cells and bodies.

Let me first explain chromosomes, genes, and DNA (figure 3.1). Each very long molecule of DNA in a cell is a *chromosome*. Each *gene* occupies its own interval along a DNA molecule. DNA is made of two

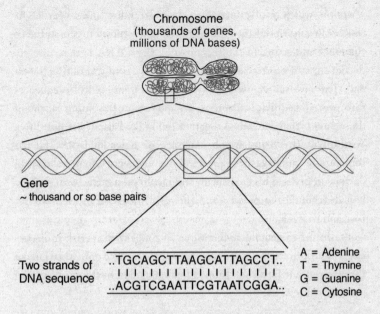

FIG. 3.1 **Chromosomes, DNA, and genes.** Chromosomes are large molecules of DNA that encode a thousand or more genes. DNA is made up of two strands of nucleotides (A, C, G, and T) held together by bonds between bases on opposite strands. Individual genes are encoded by segments of DNA sequence of varying lengths. DRAWING BY LEANNE OLDS

strands of building blocks called nucleotides. Each nucleotide contains one of four distinct *bases*; these are abbreviated A, C, G, and T. The strands of the DNA are held together by strong bonds between pairs of bases that lie on opposite strands. The number of chromosomes may be as few as just one (*E. coli*) or much greater (we have twenty-three pairs). It is the unique order of bases in a sequence of DNA (e.g., ACGTCGAATT . . .) that determines the unique information encoded within each gene.

Now let's see how information in DNA is decoded.

The first step in decoding the information in a gene is called *tran-*

scription, which involves making a single-stranded "messenger RNA" (mRNA) transcript of the gene that is complementary to one strand of the DNA molecule. In the second step, the mRNA is then directly decoded into a protein sequence in a process called translation (figure 3.2). This involves a universal genetic code to translate RNA sequences into protein sequences. Proteins are made up of building blocks of *amino acids* that are linked together in long chains. There is a direct correspondence between the sequence of bases in DNA and the sequence of amino acids in proteins. The sequence of amino acids in each protein determines their individual shape and chemical properties, whether they carry oxygen, form muscle fibers, or break down lactose and so forth.

Returning to enzyme induction in *E. coli*, the neat trick to understand is how the bacterium makes beta-galactosidase only when lactose is present. What Jacob and Monod figured out is that the production of

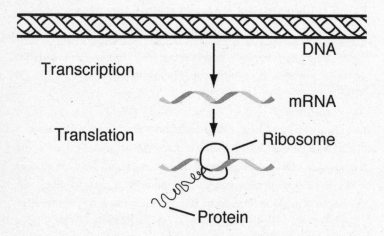

FIG. 3.2 **The decoding of information in DNA occurs in two steps.** The first step is transcription of a messenger RNA (mRNA) molecule; the second step is translation of the RNA molecule into a protein molecule. DRAWING BY JOSH KLAISS

this enzyme is controlled by a switch that resides at the beta-galactosidase gene. The switch is off when lactose is absent, but flips on when lactose is present. There are two key components of the switch, a protein called lac repressor, and the short stretch of DNA sequence near the beta-galactosidase gene to which the lac repressor protein can bind. When the repressor protein binds to this DNA sequence, the gene is off (repressed) and no RNA or protein is made. But when lactose is present, the repressor falls off the DNA, and RNA transcription and beta-galactosidase enzyme production occur (figure 3.3).

The control of enzyme production by the lac repressor is the classic

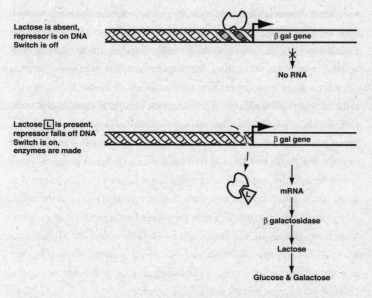

Fig. 3.3 **A genetic switch controls beta-galactosidase production and lactose metabolism in *E. coli*.** In the absence of lactose, the lac repressor binds to the switch and represses gene transcription. In the presence of lactose, the repressor falls off the switch, transcription and translation occur, and the enzyme is produced.

DRAWING BY JOSH KLAISS

example of gene logic, where a gene is used only when it is needed. There are 4288 genes in *E. coli*, and only a subset are in use at a given time. There are more than 25,000 genes in humans, and only a portion of these are used in any given cell type or organ. There are two features of gene logic in bacteria that we will see again and again, namely:

1. the regulated use of a gene occurs through the on/off binding of a DNA binding protein;
2. the DNA binding protein recognizes a specific DNA sequence near a gene.

I cannot overstate the conceptual impact of the discovery of genetic switches in bacteria. Not only were they elegant mechanisms for controlling cell physiology, it immediately occurred to Jacob and Monod that their discoveries had implications for understanding the mystery of how cell differentiation was controlled in more complex organisms, such as ourselves. They understood that the functions of blood, brain, muscle cells, etc., were all marked by the production of proteins specialized to the tasks of these tissues. Enzyme induction in bacteria was the conceptual forerunner to thinking about specialized cell functions in animals and organs. But Jacob and Monod had more talent than just genetics—they wrote beautifully and their scientific papers of the early 1960s stand as some of the most elegant, articulate, and compelling pieces in the whole of the biological literature. Their gifts were further exhibited in book-length treatments of the implications of their research. Monod's *Le Hasard et la Nécessité* (Chance and Necessity) is as well-known in literary and philosophical circles as in biology and Jacob has written several classics, including a remarkable autobiography.

So widespread were the potential implications that Monod made the quip, "What is true for *E. coli* is also true for the elephant." This was a bold leap given the state of biological knowledge at the time.

We were a long way from elephants in 1965. Was Monod right? Did the logic of this tiny bacterium extend right on up to the largest, most complex creatures on Earth? It wasn't an elephant that told us, but a tiny insect, the fruit fly, from which a series of wholly unexpected and revolutionary insights poured forth. And it started with those homeotic monsters.

The Homeobox

The homeotic mutants of fruit flies have seduced many young biologists. Flies with legs coming out of their head, or with extra pairs of wings, or feet in the place of mouthparts—part of the appeal is their B-movie character. The more erudite attraction is that these striking monsters are due to mutations at single genes that transform entire body parts into otherwise structures. How could changing just one gene change a body so dramatically? What was the "normal" job of these fascinating genes?

Finding answers to these questions depended on the development of gene-cloning technology. As these methods were becoming more widely practiced, the homeotic genes of fruit flies drew the efforts of a few brave biologists. They were aided by many years of genetic studies that revealed where on the fruit fly's third chromosome (the fly has just four chromosomes) these genes resided. Interestingly, the genes sat close together in two clusters. One cluster, the Bithorax Complex, contained three genes that affected the back half of the fly; the other, the Antennapedia Complex, contained five genes that affected the front half of the fly. Even more provocative, the relative order of the genes in these two clusters corresponded to the relative order of the body parts they affected (figure 3.4). These tantalizing but mysterious relationships raised the hope that residing in these gene complexes were some great clues to the overall logic of body patterning.

By 1983, the DNA of these two gene complexes had been isolated and was being analyzed. One of the first goals was to figure out what sorts of proteins were encoded by the 8 homeotic genes. The first

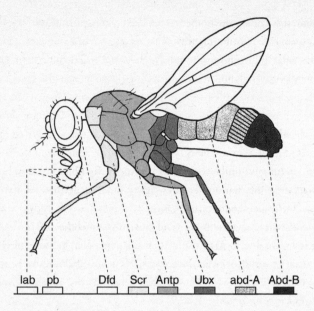

lab pb   Dfd   Scr Antp   Ubx   abd-A Abd-B

FIG. 3.4 **The *Hox* genes of a fruit fly.** Eight genes (abbreviated
lab, etc.) are located on one chromosomes in the fruit fly, and each
gene shapes the development of body regions at different positions
along the body axis (indicated by shading and stippling). DRAWING
BY LEANNE OLDS

major discovery was that of 1000 or so bases that encoded each of the
different homeotic proteins, all 8 had a short stretch of about 180 base
pairs that were very similar in sequence. Translated at the protein level,
this was a 60 amino acid domain (a part of each protein). This was
tremendously exciting because it suggested that while each homeotic
gene had specific effects on particular body regions and body parts, the
homeotic proteins shared some functional property. Molecular biologists have a tradition of naming features they find in DNA. Because the
180 base-pair sequence in homeotic genes was a small "box" of similarity in otherwise long tracts of DNA sequence, the shared DNA
sequence was dubbed the homeobox and the corresponding protein

domain it encoded, the homeodomain. The homeotic genes with these homeoboxes were subsequently called *Hox* genes for short.

But what did the homeodomain do, what property did it confer? My labmate Allen Laughon was determining the sequence of a gene in the Antennapedia Complex and wanted to know how it worked. One strategy biologists use for studying the unknown is to look for patterns in molecules, searching for similarities between newly discovered molecules and those that are better known. So Allen scrutinized the homeodomain for any similarities to other proteins that had been studied in biology, figuring that if there were any similar structures, that would give a clue to homeodomain function.

Somewhere, he had seen a structure like the homeodomain before. . . .

The lac repressor. And not only the lac repressor, but a whole catalog of DNA-binding proteins that bind to genetic switches in bacteria and yeast.

Bingo.

The similarities meant that the homeodomain was a DNA-binding domain that folded into a structure just like these other proteins. The obvious interpretation was that homeotic proteins might also regulate genetic switches in animal development and that is why they affect the formation and identity of whole structures.

This was great news but the skeptics would say we've moved from a tiny bacterium to a tiny fly. So what? What does this tell us about the majestic animals we do care about? Or ourselves?

I am familiar with that complaint. Indeed, when I went off to study fruit flies after receiving my Ph.D., some senior scientists offered their opinion that I was stepping off the edge of the Earth. Fruit flies? What would they teach us about humans or mammals? The common perception—reinforced by decades of zoology and a wide cultural divide between biologists who worked on mice, rats, or other conventional models of human biology, and those who worked on "lower" forms—was that the rules of physiology and development differed enormously between mammals and bugs or worms. So great were the differences, they believed that working on something like fruit flies was (gasp!) irrelevant.

They were in for some big surprises.

Uniting the Kingdom

Bill McGinnis and Mike Levine had no such biases about the special status of furry animals. They had also been seduced by homeotic mutants and were working in the laboratory of Professor Walter Gehring at the University of Basel in Switzerland. Once they found out that the fly's homeotic genes each had a homeobox, they took what was for them the next logical step. They purified DNA from all sorts of critters they could find around Basel or that they could beg from other labs, including various bugs, earthworms, frogs, cows, and humans, and went looking for homeoboxes.

Jackpot.

They found plenty of homeoboxes in these animals.

When the sequence of these homeoboxes were examined in detail, the similarities among species were astounding. Over the 60 amino acids of the homeodomain, some mice and frog proteins were identical to the fly sequences at up to 59 out of 60 positions. Such sequence similarity was just stunning. The evolutionary lines that led to flies and mice diverged more than 500 million years ago, before the famous Cambrian Explosion that gave rise to most animal types. No biologist had even the foggiest notion that such similarities could exist between genes of such different animals. These *Hox* genes were so important that their sequences had been preserved throughout this enormous span of animal evolution.

At first, there were different explanations for the homeobox. Some were skeptical of its significance, and suggested that maybe it encoded a mundane function, such as marking the destination of these proteins inside cells. But soon it became clear that homeoboxes offered profound insights. Jonathan Slack of Oxford University likened the discovery of the homeobox to the discovery of the Rosetta stone, which enabled the deciphering of hieroglyphs. Was the homeobox a key to deciphering the development of all animals?

A couple of years after the finding of *Hox* genes in vertebrates and other animals came another, perhaps an even bigger shock to top them all. When the arrangement of *Hox* genes was figured out in mice, they were in clusters (four in all), just as in the fly. And, even more amazing, the order of the genes in each cluster also corresponded to the order of body regions in the mouse in which they were expressed. This meant that the depth of the similarity between different animals extended not just to the sequence of the genes, but to their organization in clusters, and how they were used in embryos (figure 3.5).

It was inescapable. Clusters of *Hox* genes shaped the development of animals as different as flies and mice, and now we know that includes just about every animal in the kingdom, including humans and elephants. Not even the most ardent advocate of fruit fly research predicted the universal distribution and importance of *Hox* genes. The implications were stunning. Disparate animals were built using not just the same kinds of tools, but indeed, the very same genes!

But the surprises did not end with *Hox* genes.

More Humbling Lessons from Fruit Flies: A Tool Kit of Body-Building Genes

The body parts of fruit flies don't appear to have much in common with our own. We don't have antennae or wings, we have a single pair of movable eyes and not multifaceted bug eyes with 800 facets all staring out from a fixed position. Our blood is pumped by a four-chambered heart through a closed circulatory system of arteries and veins, not just sloshing around in our body cavity. We walk on two long legs made sturdy by solid bones, not six dainty little legs. So with such vast differ-

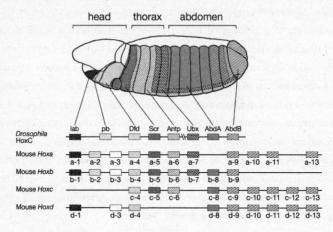

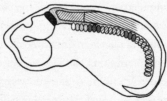

FIG. 3.5 **Clusters of *Hox* genes pattern different regions in different kinds of embryos.** Top, genes in the one cluster of *Hox* genes in the fruit fly are expressed in different regions of the fruit fly embryo. Bottom, genes in four *Hox* clusters of the mouse are expressed in different regions of the mouse embryo. DRAWING BY LEANNE OLDS

ences in anatomy, one wouldn't think that there would be anything a fly could tell us about how our organs and body parts are formed. Indeed, it had been believed that structures such as eyes had been invented from scratch as many as forty times in animal evolution to arrive at a variety of anatomical designs and optical characteristics.

In this light, the study of genes responsible for building a fruit fly eye wouldn't attract much notice. But when researchers in Walter Gehring's lab isolated the *eyeless* gene (named because of the loss of eyes in flies mutant for the gene), they discovered that the fly gene was the counterpart of a gene already known in humans. The human gene was called *Aniridia*. It was named for a mutation in humans that causes a condition where the iris in the eye (the colored portion) is reduced in size or, in severe cases, absent altogether. *Aniridia* is the same gene known as *Small eye* in mice that when mutated caused the reduction or elimination of eye formation. This discovery was intriguing and provocative because our camera-type eyes and the fly's compound eyes are very different in structure and adapted for very different needs. Why would the same gene be involved in the formation of such different kinds of eyes? Was this a fluke or a hint of something deeper?

Two additional experiments jolted the issue into focus. When the *eyeless* gene was manipulated so that it was turned on in other parts of the fly, eye tissue was induced on wings, legs, and other parts of the body (figure 3.6). This and the *eyeless* mutant effect showed that *eyeless* was a "master" gene for eye development. Without it, eye formation failed, and where it was active, tissue formed eye structures. The second experiment was to introduce the mouse *Small eye* gene into flies so that it was also turned on in weird places in the fly. What do you think happened?

The result was the same as the experiment with the fly gene—fly tissues were induced to form eye structures. However, it is important to emphasize that the tissues formed were *fly* eye structures, not mouse eye structures. So while each gene was similar and had similar effects, the final form depended upon the context of the species of the experiment, not the origin of the gene. The mouse gene induced the fly program for eye development.

The *eyeless*, *Aniridia*, and *Small eye* genes also have a much less descriptive name, *Pax-6*. The root of the name is not important, but the distribution of the *Pax-6* gene and its association with eye development throughout the animal kingdom is. *Pax-6* has been found to be

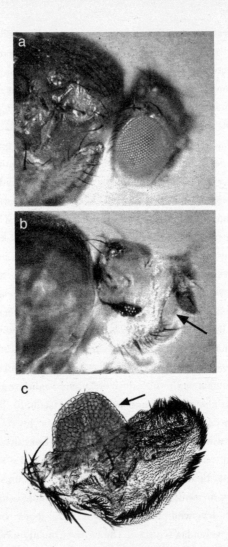

FIG. 3.6 **A master gene controls eye formation.** From top, a normal fly head with large multifaceted eye (a); *eyeless* mutant fly lacks eye tissue (b); induction of eye tissue on fly wing by induction of *eyeless* gene expression in the wing (c). PHOTOS COURTESY OF DR. GEORG HALDER, M. D. ANDERSON CANCER CENTER, HOUSTON

associated with eye formation in animals with all sorts of eyes, from simple structures such as those in flatworms, to the much more complex eyes of vertebrates. There are two alternative explanations for the widespread association of the *Pax-6* gene with eye development in the animal kingdom. This could be a remarkable coincidence in that the *Pax-6* gene was called upon repeatedly to build eyes from scratch in these different groups of animals. Or it may reflect an ancient role for *Pax-6* in the development of some type of eye in the common ancestor of all of these animals, and that role has been preserved throughout animal evolution. Before I weigh in on these alternatives, I will mention a few more surprising similarities in animal development.

In my lab, we have been very interested in the origin and evolution of appendages because legs, wings, fins, etc., are major means of adaptation for the animals that have them. A few years ago, we were studying the *Distal-less* gene (*Dll* for short), so named because when it is mutated, the distal (outer) parts of fly limbs are lost. We were curious whether this limb-building gene played a role in other species. We were pleased to find that *Dll* was deployed in the distal parts of developing butterfly limbs, and in the limbs of crustaceans, spiders, and centipedes. This told us that *Dll* plays a role throughout arthropods in limb formation (and prompted Dave Barry, the syndicated humorist, to use our paper as a rationale for why he won't eat lobsters—because they are just really big insects. Not the most artful or correct evolutionary argument, but we appreciated the attention anyway.). Since these animals are all members of one phylum and share a common jointed limb design, the use of *Dll* in all these species made sense. What we didn't expect was what we found when we and our collaborators looked at the appendages of animals that were not at all closely related to arthropods.

We found that the formation of all sorts of things that stuck out of animal bodies were associated with use of the *Dll* gene. These included chicken legs, fish fins, the appendages of marine worms (called "parapodia"), the ampullae and siphons on sea squirts, and even the tube feet on sea urchins. This was another example, like *Pax-6*, of a tool kit gene involved in building vastly different structures that only share, at most, the common feature of projecting away from the main body. These animals are also representatives of different major branches of the animal tree. So the same possible interpretations apply to *Dll* and appendage evolution as to *Pax-6* and eye evolution. Either *Dll* was tapped independently many times to build these structures from scratch, or perhaps a common ancestor used *Dll* to build some kind of projection and this role has been reused throughout animal evolution.

More similarities have been found between fly and vertebrate genes and structures. I'll mention one more example. Along the top side of the fly there is a heart that contracts to pump fluid around the inside of the body. Flies have an open circulatory system, meaning the blood that bathes tissues is not compartmentalized. It's not much of a heart by human standards, but it does the job. Geneticists have discovered a gene required for making the fly heart, and named it *tinman* (after the character in *The Wizard of Oz* who lacked a heart).

The surprise came when several mammalian versions of *tinman* were discovered, with the not quite so magical name of the NK2 family, and it was found that these genes play an important role in heart formation in vertebrates, including ourselves. Despite their great differences in cardiac anatomy and circulatory systems, flies and vertebrates have the same type of gene dedicated to the formation and patterning of their hearts.

There is one important fact to add about the Pax-6, Distal-less, and the Tinman families of proteins of flies, vertebrates, and other animals. Each of these proteins contains a homeodomain. This tells us that they are all DNA-binding proteins. The homeodomains are similar but not identical to the Hox protein homeodomains. Rather, we now understand that there are perhaps two dozen families of homeodomains. The Hox, Pax-6, Dll, and Tinman proteins belong to four distinct families. The Pax-6 proteins of different animals are more similar to one another than they are to the other families of home-odomain proteins. The Hox proteins, Dll proteins, and Tinman proteins are also more similar to other family members than to the other homeodomain proteins. The distinctions between classes of home-odomains reflect differences in functional specificity (they bind to different control sequences in DNA). Because they each bind DNA and have such dramatic effects on organ or appendage development, we know that they regulate the on/off state of genes in the developing eyes, limbs, or heart, respectively. Their large effects on development are due to their regulation of a large number of genes, their early action in the organ building process, or both (in either case, organ or body part formation collapses without them).

Rethinking Animal Evolution

The discovery that the same sets of genes control the formation and pattern of body regions and body parts with similar functions (but very different designs) in insects, vertebrates, and other animals has forced a complete rethinking of animal history, the origins of structures, and the nature of diversity. Comparative and evolutionary biologists had long assumed that different groups of animals, separated by vast amounts of evolutionary time, were constructed and had evolved by entirely different means. The connection between members of some groups—among the vertebrates, for example, or between vertebrates and other animals with a notochord—was well established. But between flies and humans, or flatworms and sea squirts . . . no way! So prevalent was this view of great evolutionary distance that in the 1960s the evolutionary biologist (and an architect of the Modern Synthesis) Ernst Mayr remarked:

> Much that has been learned about gene physiology makes it evident
> that the search for homologous genes is quite futile except in very

close relatives. If there is only one efficient solution for a certain functional demand, very different gene complexes will come up with the same solution, no matter how different the pathway by which it is achieved. The saying "Many roads lead to Rome" is as true in evolution as in daily affairs.

This view was entirely incorrect. The late Stephen Jay Gould, in his monumental work *The Structure of Evolutionary Theory*, saw the discovery of *Hox* clusters and common body-building genes as overturning a major view of the Modern Synthesis. Gould states, "The central significance of our dawning understanding of the genetics of development lies not in the simple discovery of something utterly unknown . . . but in the explicitly unexpected character of these findings, and in the revisions and extensions thus required of evolutionary theory."

Not only do homologous genes exist (which overturns perhaps the lesser of Mayr's erroneous predictions), but it appears that there are not as many roads to Rome (i.e., to an evolutionary adaptation) as once believed. The story of *Pax-6* suggests that the many types of animal eyes all took at least the *Pax-6* road. Natural selection has not forged many eyes completely from scratch; there is a common genetic ingredient to making each eye type, as well as to the many types of appendages, hearts, etc. These common genetic ingredients must date back deep in time, before there were vertebrates or arthropods, to animals that may have first used these genes to build structures with which to see, sense, eat, or move. These animals are the distant ancestors of most modern animals, including ourselves. I'll have much more to say about those ancestors and the course of animal evolution in chapter 6; before I do, I need to say more about some of the other kinds of genes in the tool kit, and I have one more fantastic set of unexpected connections to reveal.

Defining the Tool Kit

The *Hox* genes and genes for building eyes, limbs, and hearts are perhaps the most famous dozen or so master genes, but they are just a part of the collection of genes that make up the genetic tool kit for animal development. Altogether, a few hundred genes are concerned with the construction and patterning of the fruit fly. This is a small fraction of the 13,676 genes found in the fruit fly genome. The vast majority of genes have other jobs; they are involved in carrying out the routine and specialized functions of fruit fly cells.

We know about many of the tool kit genes from the same sort of approach that uncovered the first body-building genes—by isolating mutants with abnormalities. In the late 1970s and early 1980s, two geneticists in particular, Christiane Nüsslein-Volhard and Eric Wieschaus, set out to identify all the genes that were necessary to build the fruit fly larva. They found dozens of genes that were necessary to make the proper number and pattern of segments, others that were necessary for the larva to make its three tissue layers, and many more that were involved in the patterning of fine details and ornamentation of the animal. I'll say more about these various kinds of genes shortly, but the important point to make here is that Nüsslein-Volhard and Wieschaus's efforts were so systematic and thorough that they succeeded in identifying most of the genes we now know to be necessary for building the fly. In addition, many of these genes also have counterparts in vertebrates and other animals, counterparts that were discovered largely because of this pioneering work in fruit flies. Nüsslein-Volhard, Wieschaus, and Ed Lewis shared the Nobel Prize in Medicine or Physiology in 1995 for their discoveries, which paved the way for so much of embryology and, subsequently, Evo Devo.

The most striking and helpful features of Nüsslein-Volhard and Wieschaus's collection of mutants was that they had dramatic but discrete defects in embryo organization or patterning. For example, some mutants were missing entire blocks of segments while others had exactly half the normal number of segments. These genes affect the making of the basic anatomical modules—the segments that make up the insect body. In a third group of mutants, the polarity of each segment was disrupted in a regular pattern, so it was clear that these genes affected how pattern is organized within modules. In all classes of mutants, it wasn't the case that development just collapsed, but that some specific operation was disrupted while others proceeded normally.

We now know a lot about many individual genes in the tool kit. In general, all members of the tool kit shape development by affecting how other genes are turned on or off in the course of development (figure 3.7). A large portion of the tool kit is composed of *transcription factors*—proteins that bind to DNA and directly turn gene transcription on or off like the master genes I've described earlier. Another class of tool kit members belong to so-called signaling pathways. Cells communicate with one another by sending signals in the form of proteins that are exported and travel away from their source. Those proteins then bind to receptors on other cells, where they trigger a cascade of events, including changes in cell shape, migration, the beginning or cessation of cell multiplication, and the activation or repression of genes. As tissues grow, signaling between populations of cells shapes much of the local pattern within developing structures. There is a modest number of these pathways in a fruit fly (about ten) and every pathway has multiple components—signals, receptors, and various intermediaries. These components traffic the signal through compartments of the cell—from the membrane, through the cytoplasm, and into the nucleus. Mutations in any of these components can cripple signaling and disrupt development.

As biologists caught on to the idea that genes used by fruit flies were also used by vertebrates, the vertebrate counterparts of fruit fly toolkit genes were pursued every time a fly gene was newly identified. This has led to many great discoveries, and I'll close this chapter with one of the most spectacular.

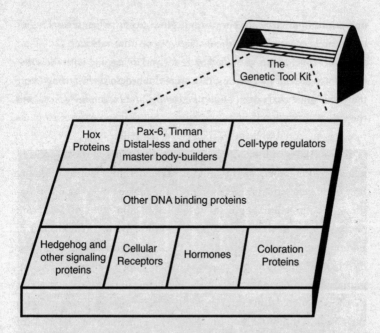

FIG. 3.7 **The tool kit of animal development.** The building and patterning of animal bodies are controlled by several different kinds of proteins in the tool kit. DRAWING BY JOSH KLAISS

From Hedgehogs to Polydactyly, Cyclops, and Cancer

Nüsslein-Volhard and Wieschaus systematically collected mutants that visibly affected the fruit fly larva and gave each gene represented by these mutants a name. Much color was added to the lore of fruit fly genetics with memorably descriptive names, many in German (the work took place in Tübingen). As such, the tool kit has genes named *knirps, Krüppel,* and *spitz,* but also *shavenbaby, buttonhead, faint little ball*, and many others. One favorite gene is *hedgehog,* named for the appearance of the mutant larva, which, like the animal, is covered with fine hairs (figure 3.8). Hedgehog is a very important molecule to many operations in the fruit fly, but its fame grew rapidly when several teams of researchers searched for the *hedgehog* gene in vertebrates.

Vertebrates have three *hedgehog* genes and in keeping with the spirit of whimsy of fruit fly genetics, they were dubbed *Sonic hedgehog* (after the video game character), *Desert hedgehog*, and *Indian hedgehog* (two

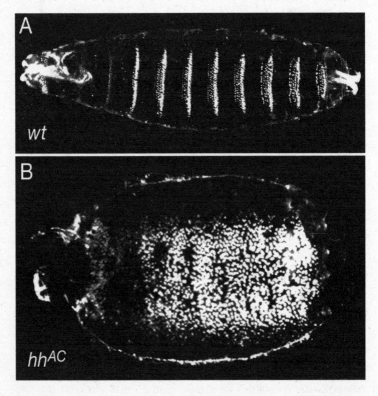

FIG. 3.8 **The cuticles of normal and *hedgehog* mutant fly larvae.** The fine hairs, or denticles, of the mutant are all bunched together and disordered, resembling the fur pattern of a hedgehog (b), in contrast to the evenly spaced segmented belts of a normal larva (a). COURTESY OF DR. BENEDICTE SANSON, UNIVERSITY OF CAMBRIDGE

legitimate animals). The first stunner came when Cliff Tabin and his colleagues at Harvard Medical School looked at how *Sonic hedgehog* was expressed in a developing chicken limb and found it was turned on in just the posterior edge of the limb bud. This was tantalizingly close to where Saunders had defined the zone of polarizing activity in his transplantation experiments decades earlier. To pursue the remarkable possibility that *Sonic hedgehog* played a role in the ZPA, the researchers used various tricks to turn *Sonic hedgehog* on elsewhere in the limb bud, and they obtained the same sort of polydactylous digit patterns as Saunders, with extra digits forming with opposite polarity. The zone of *Sonic hedgehog* expression wasn't just part of the ZPA, and the activity of the ZPA was due entirely to *Sonic hedgehog* expression. And *Sonic hedgehog* figures in more that just chicken stories.

Recall the description of humans with polydactyly? It is now clear that a type of polydactyly is due to a mutation that affects *Sonic hedgehog* expression during human limb development. That is not only a nice illustration of the homology between vertebrate limbs, but a terrific example of how a discovery in fruit flies has empowered human medical genetics.

And there is more.

Remember the cyclopic sheep (that picture is pretty hard to get out of one's mind) and the teratogen cyclopamine? We now understand that cyclopamine is an inhibitor of the Sonic hedgehog signaling pathway in mammals. It blocks part of a receptor so that cells that should respond to the Sonic hedgehog protein cannot do so. One other place where Sonic hedgehog signaling is important in vertebrate development is along the ventral midline of the developing embryo. Signaling from these "floor plate" cells is critical for the patterning of overlying tissues and their subdivision into the left and right parts of the eye field and brain hemispheres. When exposed to cyclopamine at the critical time these events would be happening, the normal steps are blocked and the newborn lamb is cyclopic. While human embryos do not get exposed to cyclopamine, it appears that ethanol can have similar affects. Fetal alcohol syndrome is a manifestation of alcohol toxicity at critical stages of human gestation and can produce holoprosencephaly. Mutations that abolish *Sonic hedgehog* gene activity or that of other components of the pathway also cause cyclopia.

Birth defects are one example of vertebrate tool kit genes gone awry. Cancer is another. Tumors develop when cells break free of their internal and external regulatory constraints, and this can be caused by abnormalities in the response to cell signaling. A pertinent example is basal cell carcinoma, the most common cancer that occurs in the skin, particularly in the face and neck of sun-exposed individuals. Cells in many of these tumors carry mutations in the gene for a receptor for Sonic hedgehog; the mutations cause too much activity of the pathway. With this insight, one approach to tumor chemotherapy would be treatment with inhibitors of the signaling pathway including—you might have guessed it— cyclopamine. (That's quite a scientific journey from stillborn lambs to plant toxins and human cancer chemotherapy.) Some brain and pancreatic cancers are now also being targeted with agents affecting the hedgehog pathway because of their association with pathway mutations.

I doubt the prospects of understanding polydactyly, cyclopia, and cancer were anywhere in Nüsslein-Volhard's and Wieschaus's minds when they started looking at mutant fly embryos in the late 1970s. But the impact of the discovery of the genetic tool kit in flies has reached farther than anyone could have possibly foreseen. And now, with a more widespread appreciation for our common genetic heritage, biologists and medical scientists routinely exploit flies and other "lower" species for clues to learn about human diseases.

The Tool Kit Paradox and the Origins of Diversity

The discovery of the shared tool kit reframes the way we have to think about the evolution of diversity. We have a lot of knowledge about tool kit genes from many animals. The distribution of the genes in the tool kit tells us that the tool kit is ancient and was in place prior to the evolution of most types of animals. We also have complete genome sequences of flies, nematode worms, a mouse, a human, a fish, and a few other animals. Comparison of genomes tell us that not only do flies and humans share a large set of developmental genes, but that mice and humans have nearly identical sets of about 25,000 genes, and that chimps and humans are almost 99 percent identical at the DNA level. The common tool kit and the great similarities among different species genomes present an apparent paradox. Since the sets of genes are so widely shared, how do differences arise? How can the same set of *Hox* genes sculpt the great diversity of arthropods? How have the great differences among all mammals—or of primates, or of apes and humans—evolved? In order to understand how different anatomies can be constructed using the same genes, we must first grasp how the anatomy of individual animals is assembled. That is a big story, which will be the focus of the next two chapters.

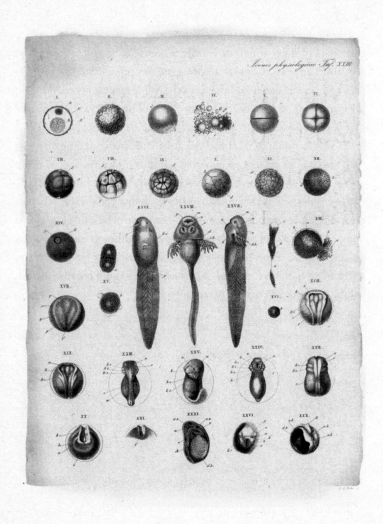

Frog development. From A. Ecker, *Icones Physiologicae: Erläuterungstafeln* (1851–1859)

4

Making Babies: 25,000 Genes, Some Assembly Required

> Hearing about something a hundred times is not as good as seeing it once.
>
> —Chinese proverb

IT'S VERY LATE on a spring night in Colorado. The lab is dead quiet. I'm repeating a procedure for the umpteenth time in the past eighteen months. After bathing the hundreds of tiny white fly embryos in this new batch of antibody I've made, I am getting nervous. Although the head of the lab, Matt Scott, doesn't know, I know that this is probably the last shot we have. I can't think of any more ways to do this experiment. I've run out of designs and tricks, so if it is a bust I'm stuck with nothing to show for a year and a half 's work. The embryos spread out on the glass slide like grains of rice and I slip them into the beam of blue light. Holy #%@*! Green stripes encircle the beautiful little maggots. Time to call the boss and run to the liquor store. We're in business.

Getting the Chicken Back

I have described the genetic tool kit for development and how its discovery was driven by the study of spectacular mutants that made the wrong number of body parts, or put a part in the wrong place, or lacked some major structure altogether. Most of the time, thankfully, nature gets it right and flies and babies are born with the right number of parts in all of the right places. But how does that happen? How do these amazing genes transform a simple egg into a complex animal?

The term "reductionism" refers to the biologist's quest to understand life's processes at the molecular level, often by breaking down—"reducing"—processes and structures into their molecular components. The approach has been enormously successful over the past half century, revealing the mechanisms of inheritance, illuminating the causes of many diseases, and creating a new $500-billion-dollar industry that provides new treatments and diagnostics in medicine. The frequent objection to reductionist thinking is that many important biological entities—cells, individuals, populations, ecological communities— are organized at levels above that of molecules, such that knowledge of molecules alone does not explain the properties of the levels above. In the same way that knowing a computer is made up of silicon, superconducting metals, and plastics does not tell us how it is constructed and functions, having an inventory of tool kit genes still leaves us well short of understanding how an animal is put together during development.

This situation is reminiscent of an earlier era, when embryologists first isolated populations of cells by brute force, hoping to learn how they become organized into tissues and organs during development. Paul Weiss once illustrated the reductionist dilemma for his fellow embryologists. He showed a picture of an intact chicken embryo, then an embryo that had been put through a blender, and finally an embryo whose blended components had been centrifuged together. Weiss bluntly stated the problem for reductionists: how to get that chicken back.

Our tool kit genes are just pieces of our genetic material, a small fraction of the 13,700 genes in a fruit fly or the 25,000 genes of a mammal. Granted, we have identified many critical pieces, but how do we get that chicken back? Or even just a fly? The challenge should ring a bell for all of us who have purchased or received a new toy or appliance in a big box, only to find bags of loose parts and an instruction sheet bearing those three deflating words "some assembly required." In the last chapter, I described how biologists went from puzzling over mutant animals to finding master genes; in this chapter I'm going to focus on the other direction, how to go from genes to building animals.

The "ahas!" in this chapter will come from making and seeing maps. In his book *Mapping the Next Millennium: The Discovery of New Geographies*, Stephen Hall described how mapmaking is one of the first stages of scientific exploration. From the great navigators of the fifteenth and sixteenth centuries to current efforts in astronomy, physics, and oceanography, scientists have sought to measure and then portray the universe, Earth, and the oceans in informative and appealing ways. Animal embryos are themselves little worlds, whose future topology is marked out by the actions of the tool kit genes.

Hall provided the apt metaphor of understanding the "geography" of the egg as a central quest of biology, and we shall make new kinds of maps to do so.

In all exploration, new instruments and technologies have played a critical role in seeing features for the first time, for both peering out into the cosmos and glimpsing the inner workings of living organisms. In embryology, the discovery of the genetic tool kit did more than identify genes for body-building. It gave us a whole new way of looking at development. By visualizing tool kit genes in action in embryos—the green stripes I first saw that night in Colorado—we can see the position and shapes of structures long before they actually form. The images of tool kit genes in embryos create a vivid, dynamic map of the geography of the growing embryo—a map that reveals to us the order and logic of how complex animals are progressively constructed from a simple egg through the work of tool kit genes.

The First Maps

The spectacle of development—from tiny egg to complex animal—is an amazing drama. In a frog, the journey from egg to tadpole takes just a few days and major events happen on the time scale of minutes to a couple of hours (figure 4.1). Within an hour or so after fertilization, the large egg, which is just a large cell, divides, or cleaves, to gave rise to 2 cells. Shortly thereafter, another cleavage, perpendicular to the first, yields 4 cells. Cleavages occur rapidly to divide the egg into 8, 16, and then 32 cells, and continue to form a ball where all of the cells are positioned toward the outer part of the sphere (surrounding a lot of nutrient yolk in the center). Then, only about nine hours after fertilization, a dramatic series of movements—*gastrulation*—begins. During gastrulation, the embryo forms its innermost (endoderm), middle (mesoderm), and outer layers, which will go on to form the tissues and organs (skin, muscle, gut, etc.) found at different depths within the

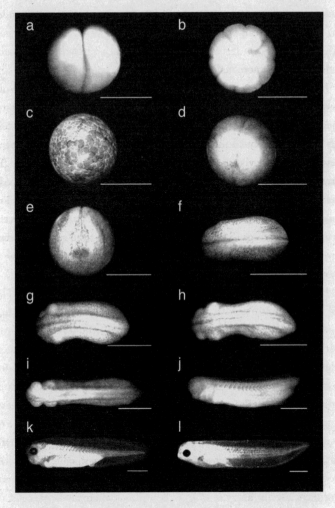

FIG. 4.1 **Development of a frog tadpole.** Progressively older embryos are shown including the first division of the egg (a), the making of the blastula (c), the formation of the inner layers of the embryo (d, e), the formation of the nervous system and somites (f–i), and the making of the eyes and their pigmentation in the tadpole (h–l). IMAGES COURTESY OF WILL GRAHAM AND BARBARA LOM, DAVIDSON COLLEGE

body. The embryo gastrulates by forming a pocket through which most of the cells originally on the outside of the embryo move inside. In just half a day or so after fertilization of the egg, the embryo is already subdivided into the three main tissue layers.

The next steps establish regions within these layers. On the topside of the embryo, a series of remarkable changes begin that will form the neural tube, the site of the future brain and spinal cord. After just one day of development, folds and bulges mark future regions where the head, eyes, and tail will develop. The organs and appendages of the tadpole then begin to form. The dorsal fin forms on the second day and the eyes become pigmented, the heart and vascular system develop and red blood cells are evident by early on the third day. The free-swimming tadpole will remain fully aquatic while it undergoes further development of its limbs, its tail is reabsorbed, and eventually it assumes the adult form.

The making of a fly larva is also a rush job (figure 4.2). The oblong fly egg starts with a single fertilized nucleus and in just a few hours there are 6000 cells forming a solid sheet around the yolky interior. The embryo then undergoes gastrulation, forming the inner, middle, and outer layers. The main trunk of the embryo then begins to extend and the making of grooves rapidly sculpts the segmental pattern of the embryo. Within this half-day-old embryo, various organs of the larva form and pockets of cells that will give rise to future structures of the adult are set aside. In just one day after fertilization of the egg, a ravenous, mobile larva hatches. This animal will grow rapidly, molt twice, form a pupa, and then emerge as a fly about nine days later.

Frog and fly embryos and larvae are very vulnerable to predators. The sprint to complete development is a survival imperative and of the hundreds of eggs produced by a female, only a fraction will reach adulthood. Humans have a different ecology, and our development takes place in maximum security and unfolds, at least initially, at a much slower pace. The initial divisions of the fertilized human egg

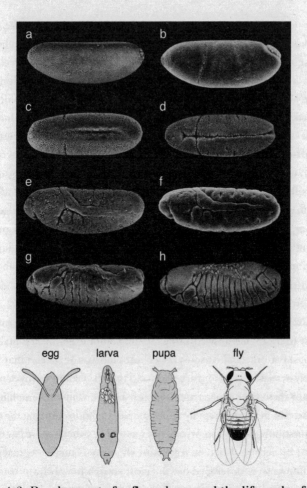

Fig. 4.2 **Development of a fly embryo, and the life cycle of a fly.** Top, electron micrograph pictures of progression of fly development, including formation of inner layers of the embryo (a–d) and the making of body segments (e–h). The events of a–h transpire in less than twelve hours. Bottom, after the embryo hatches into a larva, it grows and undergoes several molts before forming a pupa; the adult emerges from the pupal case several days later. IMAGES COURTESY OF RUDY TURNER, INDIANA UNIVERSITY; DRAWINGS BY LEANNE OLDS

occur about every twenty hours, so it takes about as long to make a 32-cell human embryo as to make a complete frog tadpole. Gastrulation doesn't occur until about day 13, and it takes about three weeks to form a distinct head region. Paired segmental bulges form along its back, which will distinguish it as a vertebrate (these *somites* will give rise to the vertebrae and surrounding muscle and skin). The human embryo is about 2.5 millimeters in length at this time and it will be another eight (long!) months before birth.

Watching an embryo develop, we start asking several obvious questions. How does the embryo know which part will be the head and which will be the tail? Or the top or bottom? How does it decide where to put the eyes, legs, or wings? If we think a little more about the future potential of that initial fertilized egg cell, which will go on to form muscles, nerves, blood, bone, skin, liver, etc., we might ask how all of these potentials are realized. At what point in embryonic development is a cell's fate sealed?

Fascinating questions like these prompted the great pioneers in embryology to try to find answers using the simplest of experimental manipulations. For practical reasons, they chose embryos of species that were widely available, easy to work with, and that they could watch develop. Usually these were aquatic animals, such as sea urchins and amphibians, whose eggs developed rapidly under simple conditions. Among the earliest questions asked by embryologists was: What structures are later produced by cells in different regions of the early embryo? A variety of techniques were developed but the most straightforward was to mark individual cells with a harmless chemical dye and then to see where that cell and all of its daughters wound up. This led embryologists to construct maps—so-called fate maps—of early embryos that revealed the relative position of the cells that give rise to particular structures.

From fate-mapping experiments, atlases of the embryos of many animals have been drawn. Analogous to the longitudes and latitudes on the globe, coordinates of embryos were identified that defined the

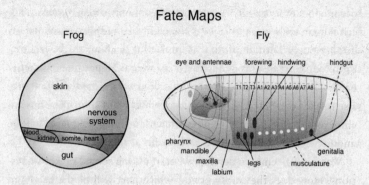

FIG. 4.3 **Fate maps.** The fate maps of an early frog and an early fly embryo. The future body parts that will develop from different regions are indicated. DRAWING BY LEANNE OLDS

future position of tissues, organs, and appendages. Two maps, of the frog and the fly embryo, are shown in figure 4.3 (two different techniques are used, but the idea is the same). We can see that in a frog embryo, the epidermis, nervous system, blood-forming tissue, heart, nervous system, skin, and gut come from particular latitudes and longitudes of the early embryo.

The fly embryo is a different shape than the sphere of the frog egg, being oblong like a rugby ball or an American football. The various parts of the future adult fly also come from discrete coordinates in the map, and we can see from the map that structures that will form at different positions along the body axes of the fly originate from different positions along the body axes of the early embryo.

The Geography of the Embryo

Fate maps reveal that, by some point in development, cells "know" where they are in an embryo and to what tissues or structures they belong. In the terms of our geography analogy, cells, tissues, and organs have a specific position on the globe of the embryo defined by their longitude, latitude, altitude (if projecting out from the body), and depth (within the layers of the body), as well as a "national" identity (nerve cells, liver cells, etc.). All cells are descended from the initial single fertilized egg cell. It is obvious that a whole lot of information must be generated in the course of development to create unique addresses for dozens of cell types, tissues, and organs at specific positions in the embryo. How do cells "learn" their position and identity? This is the collective work of the tool kit genes. And there is a sensible logic to the order of tool kit gene actions such that positions in the embryo are defined on a progressively finer scale.

Before I show pictures of how the genes actually do the job, I illustrate the general logic of embryo geography in figure 4.4. Look carefully at this figure before reading on. The key idea is to picture embryos as globes upon which coordinates are progressively determined and refined in several steps.

Now, let's see how tool kit genes actually do this.

The Shapes of Things to Come:
Tool Kit Genes and the Drawing of Bands,
Stripes, Lines, Spots, Dots, and Curves

The geometry of the embryo's coordinate systems, with its parallel and intersecting lines of longitude and latitude, imposes some spatial order on how the program of tool-kit genes unfolds. This geometry is also reflected by the physical contours of developing embryos, which are sculpted by periodic grooves, form smooth curves, and have many spherical features. The populations of cells that make up the major subdivisions of embryos, or the positions of developing organs or other specialized structures, are often first marked as simple geometrical shapes—bands, stripes, lines, spots, dots, or curves—of tool kit gene expression. Francis Crick, the Nobel laureate and codiscoverer of the structure of DNA, once remarked that "embryos are very fond of stripes." Indeed, but these stripes and other shapes are much more than aesthetically pleasing pictures of the tool kit genes in action as embryos develop; they reflect the basic operations through which the complex architecture of animals is progressively built up from geometrically simple patterns.

Whether building a tiny fruit fly or a large mammal, the general logic of tool kit genes' action in organizing, subdividing, and specifying and sculpting parts of the embryo becomes clear when visualized. Seeing each gene's job in marking out the geography of the embryo will help us make sense of a complex process as the product of many simpler individual operations. The complexity of the animal arises from many operations taking place at the same time and in succession during development. I can't possibly describe or show every detail of the development of an entire animal at the level of genes in a single chapter. Nor need I do so in order to get the picture across. I am going to illustrate development here in a few broad strokes. By focusing on the steps that shape the major features of animals, we'll see quite vividly how the future forms of animals are laid out. The color plates illustrating this process are just a sample of tens of thousands of pictures gathered by researchers over the past two decades. They are embryology's equivalent of satellite photographs of the Earth. I'll begin with the geography of the fruit fly.

The Making of a Fly

Viewed in ordinary light, the freshly laid fly egg belies none of the dramatic events unfolding within it. Tool kit genes, stirred into action by fertilization, are starting to mark out the geography of the developing embryo. While all cells in the growing embryo contain the same DNA (and the same genes), the tool kit genes become active

1. Define the poles

To establish a coordinate system, the poles of the embryo are defined first. Every embryo has North (top) and South (bottom), West (head) and East (tail) poles. The two major axes of the embryo connect these poles.

2. Subdivide the major axes into finer divisions

The East-West and North-South axes are subdivided into regions of longitude and latitude, respectively. Initially, these subdivisions are large, perhaps only demarcating Northern and Southern hemispheres or Eastern, Middle, and Western regions.

3. Refine intervals into series of modules

The lines of longitude and latitude are made progressively finer, from 90° to 30° to 15° and so forth. In many embryos, particular lines of longitude define the anatomical modules that are the basic building blocks of that animal's design.

4. Define identities of different modules

The initially similar modules are distinguished by their position along the East-West axis. A system of discrete longitude interval is superimposed.

5. At specific coordinates of longitude and latitude, new "worlds" within the world form.

The positions of future organs and appendages longitude are very important to the architecture of the animal. These positions are specified by a combination of longitude and latitude. Initially, small groups of cells at particular coordinates (e.g. 30° W, 10° s) are recruited to form specific body structures. These small groups of cells must often multiply dramatically before the organ or appendage reaches its final size. These structures are themselves often modular. In order to sculpt the form of these structures, the processes of 1-4 above are repeated within these growing worlds such that:

FIG. 4.4 **General logic of embryo geography.**

5a. Axes are defined

The North-South and West-East axes of organs and appendages are usually established early as the future structure is beginning to be distinct from surrounding cells.

5b. Major axes are divided into intervals

Both axes are often subdivided into intervals.

5c. The third axis is formed and intervals are refined into modules

Many structures also begin to elaborate a third axis perpendicular to the first two axes to become 3-dimensional. The organ or appendage, such as the segment of a limb Some intervals may form the basic modular architecture of the organ or appendage, such as the parts of a limb.

5d. The initially similar modules are distinguished by their position along the East-West axis and develop different size, shape, or architecture from one another.

6. Detailed patterns of modules are ornamented, sculpted, and colored

The coordinate systems of anatomical modules are often sufficiently refined to specify positions nearly to the precision of rows, groups, or even single cells. The fine details of form including shape, color, and the locations of specialized cell types—for sensory input, defense, decoration, etc.—is often assigned at the level of small populations of cells in structures that may be made up millions or tens of millions of cells.

only in parts of the embryo and only at particular times in development. We can see their on/off patterns with powerful technologies that light up their RNA or protein products within embryos and developing body parts. These patterns reflect the order and logic of the making of the animal.

Longitudes, the East-West Axis

Within a couple of hours after fertilization, the fly embryo is about 100 cells long from west to east. A small number of tool kit genes mark out the western, middle, and eastern regions of the embryo in bands of about 15–25 cells (plate 4a), some of which overlap.

These subdivisions are transient, but before they fade another tier of tool kit genes turns on in sets of seven stripes over the eastern two-thirds of the embryo. These stripes are 3–4 cells wide and are separated by interstripes of 4–5 cells (plate 4b). Each stripe and interstripe together cover a pair of future segments; this group of tool kit genes is dubbed the pair-rule genes.

These stripes are also transient. Just as these beautiful, regular stripes start to fade, another group of genes turns on in patterns of fourteen lines—some 1–2 cells wide, others a bit wider—over the eastern two-thirds of the embryo (plate 4c). The future larva will have fourteen main segments, so the regular intervals of these stripes corresponds to one stripe per future segment. Most of these fourteen stripe patterns persist throughout development, and within a few hours of their appearance, the physical segmentation of the embryo occurs. Some of these stripes precisely mark the boundaries between segments, others mark different sets of longitudes in the middle of each segment.

As the segmental modules are laid out in the embryo by these tiers of genes acting in succession, a fourth group of genes is activated that distinguishes the identities of the modules at different longitudes along the east-west axis. These are the *Hox* genes whose realms generally span from two to about seven segments and whose patterns persist throughout development (plate 4d). The *Hox* genes will determine what does or does not take place within individual or sets of segments.

Latitudes, the North-South Axis

At the same time the east-west axis is being subdivided, the north-south axis is also being carved into latitudes by a different set of tool kit genes. Much like the first group of longitude genes, the first tier of latitude genes mark out broad regions of the northern, equatorial, and southern zones of the embryo (plate 4e).

The latitudes of the embryo do not correspond to repeated modules, but some delineate the future tissue layers of the animal. For example, all of the cells that express the gene shown in plate 4e will wind up on the inside of the embryo through the process of gastrulation and form the animal's middle layer, or mesoderm, which will form the musculature and other tissues. Cells just north of these, originating near the equator, will be pulled south and form the epidermis on the underside of the animal and the nerve cord.

Worlds Within a World, Tool Kit Genes Mark the Spots of Future Organs and Appendages

When longitudes and latitudes are defined and refined, the information exists to specify positions on the embryo with respect to both axes. This is how organs and other structures are placed, and where the master genes for organ building come into play. And, depending on the number of each kind of particular structure to be formed, one or several pairs of coordinates may mark the site of construction.

For example, every fly will have three pairs of legs on the thorax, one pair per thoracic segment. In a developing embryo, the master gene for limb-building, the *Distal-less* gene (recall chapter 3) is turned on in several places, toward the southern end of three segments just west of the middle of the embryo (plate 4f, bottom). Notice that Dll is not activated in the eastern segments. That is because here the Hox proteins govern what will happen in each segment and Dll is prevented from turning on any farther east (these will be the abdominal segments, and they will not bear legs).

Similarly, the fly will have two pairs of wings and we can see that a master gene involved in wing formation is cued to be active in cells just north of where the *Dll* gene marks the future legs in the second and third thoracic segment, but not anywhere farther east (plate 4f, top). The relative position in the developing wings foreshadows their relative position in the adult fly (wings on top, legs underneath).

The future legs and wings are tiny at this stage. We can count perhaps 15–20 cells in each. But they will grow a thousand-fold or more in a few days, and become larger than the embryo was at the time they first developed. These structures will eventually be well organized into discrete parts. That organization depends upon the elaboration of a system of coordinates within the growing leg, wing, or organ. This system is set up when the body part is still small, when cells inherit the information about where they are in a segment, and it is refined as the structure grows. For example, a small cluster of 20 wing cells will grow into 50,000 cells with the west (front), east (back), north (top), and south (bottom) parts marked out by tool kit genes (plate 4g). The lines of longitude and latitude within these structures serve both as physical boundaries (such as the edge of the wing) and as reference points or landmarks around which further subdivisions are organized.

The coordinate system within a growing appendage is sufficiently refined to assign the position and identity of rows, clusters, and even some individual cells. For example, two obvious features of the fly wing are the sets of veins that act as structural struts during the very rapid beat of the wings in flight and the rows of sensory bristles at the leading edge of the wing. The position of these veins and the spaces between the veins are marked out by tool kit genes long before the veins actually form, about a week before the bug actually flies. At the leading edge of the wing, rows of bristles form that are positioned with respect to both axes in the wing such that they develop on both sides of the equator, but not in the eastern half of the wing. The genes that build bristles are activated in these positions long before the bristles are visible (plate 4h).

Making Serial Repeated Modules Different

The two pairs of wings in a fly are very different in form and function. The front forewings are large, flat, venated, and powered for flight. The rear hindwings are much smaller, balloon-shaped, without veins, and serve to balance the fly by sensing and correcting yaw, pitch, and roll during flight (flies without hindwings crash to the ground). The hindwings start development in a similar fashion as the forewings; the coordinate systems of the two wing types, visualized as patterns of gene expression, are identical (plate 4i). But the hindwing winds up being a very different size, shape, and pattern than the forewing.

The key to making these two wings different, which develops in two adjacent segments (i.e., at different longitudes), is a *Hox* gene known as *Ultrabithorax* (*Ubx*). This *Hox* gene is activated in all cells of the hindwing, and not in the forewing (plate 4j). The action of *Ubx* modifies the developmental program within the hindwing such that a subset of wing-patterning genes is suppressed, while other genes are used in unique ways. For example, none of the venation genes are activated in the hindwing and the bristle genes are not activated along its leading edge. The differentiation of the hindwing from the forewing by the *Ubx* gene illustrates how a fundamental feature of animal design, the differentiation of serially repeated parts, is controlled by *Hox* genes at specific longitudes along the main body axes. The same principle we see in the fly applies to the making of our body plans and body parts.

The Making of a Vertebrate

While the eggs of vertebrates differ greatly in size and character, from tiny mammal eggs to the enormous shelled eggs of large birds and reptiles, and the adult forms range from guppies to elephants and dinosaurs, all vertebrate embryos pass through a stage of development where they appear somewhat similar. This is when the main east-west body axis (head-tail) has formed and the different tissue layers are well-defined (north-south) such that the neural tube and notochord (a stiff rod of cells along the back of all vertebrates) are evident, and the regular, paired bumps of somites form a pattern of repeated modules along most of the length of the animal.

I will focus first on the steps leading to the formation of this stage in order to illustrate the formation of the basic design of the vertebrate body plan. These steps include the establishment of the major body axis, the subdivision of the brain, and the formation of the somites that will give rise to the vertebrae, ribs, and other modular elements of the main body axis. Then, I will zoom in on the development of the limbs to illustrate how more detailed patterns are sculpted. I will draw on studies of frogs, fish, mice, and chicks to paint a general picture. The principles are very similar in each species but some events are better understood or easier to see in certain animals. The important thing is to get an overall picture of the building of the vertebrate body plan; we won't worry about differences in details.

Axis Formation and Generation of Tissue Layers

Much of what we know about axis formation and the generation of the three primary tissue layers in vertebrates was first discovered in frogs. Amphibians in general have very large eggs that are deposited in great numbers; this makes experiments with them much easier than with mammalian eggs, which are tiny, produced in small numbers, and develop inside the mother. While all vertebrate embryos reach a similar organization after gastrulation, how they get there is a bit different because of differences in the proportion of cells and yolk in the early embryo. Even though the early steps in a frog embryo don't look the same as in a mouse embryo, the set of tool kit genes that organize their axes and tissue layers are generally similar.

The axes and tissue layers of vertebrate embryos are organized by a chain of inductive events, where the production of one molecule induces others and so on. The formation of the west-east axis follows that of the north-south axis. One of several key molecules is the tool kit protein Chordin. It is produced by cells around the dorsal lip of the blastopore (plate 4k), the region Spemann and his student Hilde Mangold showed had the ability to organize the north-south axis. Other proteins organize the head-to-tail axis, such as the Frzb protein, which is expressed in cells toward the future head of the embryo (plate 4l).

Subdivisions of the Brain

The neural tube is one of the first overt regions to form in the embryo. This tube will give rise to the future brain and spinal cord. The brain will be subdivided into three primary regions (the forebrain, midbrain, and hindbrain); these will become further subdivided into parts that are specialized for various functions, ranging from smell and vision to the reflex control of involuntary activities such as respiration and heartbeat. Before these subdivisions are evident, and long before their functions are established and integrated, tool kit genes mark out the regions of the neural tube that are fated to become parts of the brain. For example, the expression of one tool kit gene marks the future forebrain and midbrain, while a second gene marks out the hindbrain and the midbrain/hindbrain boundary. The cerebellum forms just east of this boundary.

Other tool kit genes turn on in stripes that mark the future positions and boundaries of the rhombomeres, the seven subdivisions of the hindbrain in all vertebrates. Some stripes span pairs of adjacent rhombomeres, others mark out alternate pairs of rhombomeres (plate 4m). The different modules of the hindbrain are then staked out by the staggered expression of *Hox* genes. Recall that there are four clusters of *Hox* genes in most vertebrates. Clusters are denoted with letters (a–d) and the gene in a cluster with a number (1–13). In the hindbrain the expression of adjacent *Hox* genes mark out unique and overlapping sets of rhombomeres, as the five components of plate 4n show. Specifically, *Hoxa2* is expressed in rhombomeres (r) r2–r4, *Hoxb2* is expressed in r3–r4, *Hoxb1* is expressed in r4, *Hoxb3* is expressed in r5 and r6, and *Hoxb4* in r7 and zones farther east in the spinal cord. These five genes are sufficient to create a unique *Hox* "code" for each rhombomere from r2 to r7. Other genes distinguish r1 (the future cerebellum) from other rhombomeres. This same logic of making a series of initially similar modules and then making them different from one another applies to another prominent feature of vertebrates, the making and differentiation of the segmental organization of the body plan.

Segmentation of the Vertebrate Embryo One Somite at a Time

Somites are the building blocks of vertebrate bodies. They give rise to the modular parts of the vertebral column, associated ribs, and muscle groups. They appear in embryos as evenly spaced paired seg-mental bulges along the main body axis and form one at a time from head to tail (west to east) in all species. There is an orderly pace of somite formation, occurring about every twenty minutes in a zebra fish embryo, every hour and a half in a chicken embryo, and every two hours in a mouse embryo. Humans form about forty-two somites, mice about sixty-five, and snakes up to several hundred.

The clocklike precision and head-to-tail progression of somite for-mation is foreshadowed by the expression of several tool kit genes. A handful of genes is expressed in the unsegmented part of the embryo before somite formation, and their expression oscillates during each round of somite production. At the front of this domain, separate stripes of tool kit expression appear that mark the boundary of the newly forming somite and, farther toward the head, stable stripes mark the boundaries of somites made earlier. Over the course of embryo development, snapshots of tool kit gene expression reveal the stepwise progression of somite formation (plate 4o).

The somites are initially identical in appearance, but they will give rise to distinct types of vertebrae, ribs, and musculature, depending upon their position along the head-to-tail axis. The distinct identities and fates of somites are foreshadowed by the expression patterns of *Hox* genes along the main (west-to-east) body axis. *Hox* genes turn on with distinct western (anterior) boundaries at the level of particular somites, with expression generally trailing into more eastern (poste-rior) zones (plate 4p).

The staggered boundaries of *Hox* gene expression result in unique combinations of *Hox* genes being expressed in different somites. Furthermore, the sharp anterior boundaries of individual *Hox* domains often mark the boundaries between different types of struc-tures in the vertebrate skeleton. For example, the boundary of *Hoxc6* expression marks the boundary between cervical and thoracic verte-brae in all vertebrates.

The Making of a Limb: The Leg Bone
Is Connected to the . . .

Once the basic body plan is laid out and the repeating pattern of
somites is far along in its development, positions in the embryo begin
to be marked where various organs and appendages will form. Tool kit
genes for building three-dimensional structures are activated, and body
part construction begins.

Just like us, mice and other vertebrates possess many organs. I shall
focus here solely on the limbs. The four-limbed pattern of vertebrates
is ancient and there are many general similarities to limb development
among all vertebrates. The mouse and the chicken have been studied in
the greatest detail and I'll rely mainly on these two animals to paint a
picture of how these marvelous pieces of anatomy are constructed.

The limbs begin as tiny buds that grow out of the flank of the
embryo at two specific coordinates along the east-west axis. The forelimbs arise at different somite numbers in different vertebrates, but are
always found at the cervical/thoracic boundary. The westernmost bud
will become the forelimb (a leg in mice, a wing in chickens), the east-
ernmost bud the hindlimb (legs in both species). At a very early stage,
we can see one tool kit gene whose expression marks the position of
the tiny pads of limb tissues just as they begin to form (plate 4q).

The buds, while initially very small, are three-dimensional and possess three axes along which the top (back) and bottom (palm), front
(thumb) and back (pinkie), and proximal (e.g., shoulder) and distal
elements (digits) of the limb will form as the bud grows dramatically.
Specific tool kit genes organize these axes in the early limb bud. For
example, the Sonic Hedgehog signaling molecule is made in a zone in
the most posterior part of the limb bud (plate 4q), the FGF8 signaling
protein is made around the entire outer ridge of the limb bud (plate
4r), and the *Lmx* gene is activated only in cells in the top half of the
limb (plate 4s).

Other tool kit genes turn on in patterns that foreshadow the physical development of the long bones, digits, joints, muscles, and tendons of the mature limb. The development of these elements proceeds in a proximal-to-distal order, with the future position of the upper arm or thigh, forearm or calf, and hand or foot specified in sequence. Bone develops from condensations of cells that first form a cartilage template, which is replaced by bone. The patterning of the limb is visible first as cartilage patterns. But even before these patterns are visible at the cellular level, the expression of the tool kit gene *Sox9* prefigures the pattern of condensations (plate 4t). Joints arise in zones between these condensations and even before these spaces are visible at the cellular level, the striped expression of the *GDF5* gene marks the future position of the shoulder, elbow, wrist, and joints between the hand and finger bones in the forelimb and the positions of the knee, ankle, and foot and toe joints in the hindlimb (plate 4u). The future position of tendons that attach muscles to bones of the limb is prefigured by expression of *scleraxis*, another tool kit gene (plate 4v).

The beauty of the limb is also sculpted by death. The separation of digits in mice, chicks, and humans is due to the death of the tissue between the digits in the developing limb. Within the padlike hands and feet, these interdigital zones are marked by expression of different tool kit genes that instruct the cells in the zones to undergo programmed cell death (plate 4w). In a cookie-cutter-like fashion, the interdigital tissue is carved away, leaving the digits. Interestingly, in ducks, an additional tool kit member is expressed in the interdigital zones to block the death-promoting signal—and this leaves the interdigital webbing of duck feet.

While all limbs are composed of the same structures—bones, tendons, muscle, joints, etc.—these structures differ in size, shape, and number within the limb. For example, the upper arm is composed of one long bone, the forearm contains two long bones, and the hand contains up to five digits. The size, shape, and number of limb elements are influenced by a subset of *Hox* genes (primarily the

Hoxa9–13 and *Hoxd9–13* genes) that are expressed in complex, partially overlapping patterns throughout the development of both the forelimb and hindlimb. In most species, the forelimbs and hindlimbs are also constructed differently. Our arms and legs, hands and feet, fingers and toes are different forms of the same structures. In other animals, such as birds, kangaroos, or *T. rex*, the differences between forelimbs and hindlimbs can be more dramatic. A set of tool kit genes is selectively expressed in either the forelimb or hindlimb and governs the development of differences between these serially homologous appendages.

Finishing Touches: Order on a Fine Scale

One of the most striking truths about animal body patterns is their regularity at all scales, from the overall body plan to the fine details of an individual structure or body part. The tiling scales on a butterfly wing and the regular spacing of feathers on a bird are good examples of the latter. While the location of cells can be specified fairly precisely, this is not the only means of achieving regularity of patterns. The

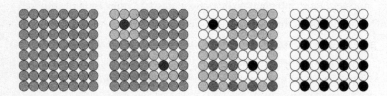

FIG. 4.5 **The generation of regular spacing patterns.** In an initially uniform field of cells (first panel), two cells begin to differentiate (black circles, second panel) and inhibit cells in contact with them from doing so. Cells in other regions begin to differentiate and inhibit their nearest neighbors (third panel), which eventually establishes a regularly spaced pattern of cells (last panel). These cells now may form bristles, feathers, or other structures.
DRAWING BY JOSH KLAISS

spacing of many individual elements in a larger array is often accomplished by a process dubbed *lateral inhibition*. The principle is simple, the effect is beautiful.

Imagine a cluster of people scrunched together. Each person is given the instruction to move one arm length away from all others in every direction. In effect, each person creates an arm's length zone of inhibition around himself or herself such that no other individual is in that zone. The result will be an array of evenly spaced individuals (so long as arms are the same length, which they are in figure 4.5).

Cells do the same in generating order on a fine scale. The general mechanism is for cells that are to become certain types of structures to create a local zone of inhibition around themselves. Only cells outside the effect of the zone are able to also become that structure. The net effect is a regular pattern—of hairs on insect bodies; feather, scale, and fur patterns on birds, reptiles, and mammals; and the beautifully packed compound eyes of arthropods. All of these patterns are generated locally by cell interactions, not specified by global coordinates. In embryos, the establishment of these patterns appears as regularly spaced patterns of expression of genes involved in the further development of a structure. For example, *Sonic hedgehog* is expressed very late in chicken development in each of the future feather buds before they actually develop (plate 4x).

Complexity from Simplicity: Seeing the Invisible

François Jacob has pointed out that all of our explanatory systems, whether mythic, magic, or scientific, share a common principle. They all seek, in the words of physicist Jean Perrin, "to explain the complicated visible by some simple invisible." I would argue that the revolution in understanding how animals develop came about because we were able to go one step further—by making the "simple invisible"

visible. The ability to see stripes, spots, bands, lines, and other patterns of tool kit gene expression that precisely prefigured the organization of embryos into segments, organs, and other body parts provided many "Eureka!" moments when the role of a gene in a long studied process became exquisitely clear. Stripes that foreshadowed segments, patches that revealed powerful zones of organizing activity, and other patterns that marked positions of bones, joints, muscles, organs, limbs, etc.—all of these connected invisible genes to the making of visible forms.

Furthermore, the revealed order of tool kit action in animal development made logical sense. Just as in the construction of a building, where there is an order to the sequence of steps—the foundation is poured, the supporting walls and beams erected, the floors laid, major ducts placed, plumbing, electricity, drywall installed, etc.—there is an order to building animals, from the making of the basic body plan to the fine detailing of individual body parts. And, from the logic of this order, we then understand how monstrosities result when the operation of a tool kit gene is damaged by mutation. When a step is omitted, all dependent steps are abnormal.

I have said that the role of an individual tool kit gene is easy to understand when visualized in action and I have shown many geometrically simple patterns to make my case. But the building of an entire animal *is* complicated. The complexity arises from the parallel and sequential action of tool kit genes—dozens of genes acting at the same time and place, many more genes acting in different places at the same time, and hundreds of tool kit genes acting in sequence as development progresses. The chain of parallel and successive operations is what builds complexity.

I have glossed over the questions that by now may be taking shape as you contemplate tool kit gene patterns and think about the chain of tool-kit genes: What connects the links of the chain? How do tool kit genes know in what order to act, or where to act in the embryo or body part?

The making of an animal involves one more set of genetic invisibles—little devices in the DNA that govern where and when genes are activated. In the next chapter, I will describe the fantastic little devices in the genome that draw the beautiful patterns of gene expression you've seen here and that are the key links between the chains of tool kit genes that build animal complexity and diversity.

Dark matter in the universe (top) and genome (bottom). The top
is an image of the galaxy cluster CL0024+1654; the dark matter
appears as hazy cloud in the center. The bottom image is of a
microarray of the fruit fly genome—the bright spots are DNA
that encodes genes, the darks spots are DNA that is not expressed.
UNIVERSE COURTESY OF EUROPEAN SPACE AGENCY, NASA, AND JEAN-PAUL
KREIB (OBSERVATION MIDI-PYRENEES, FRANCE/CALTECH, USA).
GENOME COURTESY OF DR. TOM GINGERAS AND AFFYMETRIX, INC.

5

The Dark Matter of the Genome: Operating Instructions for the Tool Kit

It is indeed a feeble light that reaches us from the starry sky. But what would human thought have achieved if we could not see the stars?

—Jean Perrin

CONSIDER FOR A MOMENT development from the egg's perspective—what a job lies ahead of all those cell divisions and movements, and of making tissue layers, segments, and body parts. We have seen that development proceeds in logical steps, but where are the instructions for each step? How is it that wide stripes are made before narrow ones, or some bones positioned before others? Or how is that some bones are long and thin, and others short and fat? How do the tool kit genes know where to act and when, to shape the development of form? Where are the operating instructions for the tool kit?

To answer these questions, I am going to do two dangerous things. First, I am going to call upon cosmology for an analogy. This is risky because I know very little about the study of the universe, but I have learned that there is a handy analogy between the makeup of the universe and the structure of the genome. And second, I am then going to mix this analogy with another. I will justify committing this writing misdemeanor on the grounds that this chapter presents some of the most challenging, yet also the most revealing and conceptually important information in this book. Bear with me.

For most of its history, astronomy has been about what could be seen in the sky, first with the naked eye and then with ever more powerful telescopes. But while much of what can be seen is always becoming better understood—the formation of stars, the structure of galaxies, and the collapse of suns—cosmologists have only recently confronted the prospect that only a small fraction of the matter in the universe is visible (emitting light or radio waves). The behavior of some visible objects such as galaxies is affected by more abundant, invisible "dark matter" and "dark energy."

The analogy with genetics is that for decades, because of the simplicity of the genetic code, we biologists have been able to see the "stars" of the genome, to see exactly where genes are encoded in DNA. But we too now appreciate that in most animals' genomes, the genes that we see occupy just a small fraction of DNA. A much larger part of our DNA consists of sequences that are not part of the simple code for any gene and whose function cannot be deciphered simply by reading the sequence. This is the "dark matter" of the genome. Just as dark matter in the universe governs the behavior of visible bodies, the dark matter in our DNA controls where and when genes are used in development.

This chapter is all about the dark matter in DNA and how, by virtue of controlling how tool kit genes are used, it contains the instructions for making and patterning body parts. These instructions are embedded in the dark DNA as *genetic switches* (my second analogy). You may not have heard about switches before this book. They have not received nearly as much attention, either in the lab or in the press, as they deserve. But this is more a reflection of the challenge biologists have had in finding them and deciphering how they work, not of their importance. Molecular biologists have only relatively recently been able to peer into the dark and reveal the location and properties of switches. The most surprising and crucial feature of genetic switches is their ability to control very fine details of individual tool kit gene action and anatomy. The anatomy of animal bodies is really encoded and built—piece by piece, stripe by stripe, bone by bone—by constellations of switches distributed all over the genome.

Switches are key actors in both dramas here—development and evolution. These switches draw the beautiful patterns of gene expression we saw in the last chapter. It is the switches that encode instructions unique to individual species and that enable different animals to be made using essentially the same tool kit. And switches are hotspots of evolution—they are the real source of Kipling's delight—the makers of spots, stripes, bumps, and the like. Part genetic computer, part artist, these fantastic devices translate embryo geography into genetic instructions for making three-dimensional form.

Seeing in the Dark

In cosmology, biology, as well as other sciences, the existence of particular entities is detected either directly, by observation, or indirectly, by observing the effects on other entities that are more easily visualized or measured. The evidence for dark matter in the universe is all indirect, based upon observation of the velocities and rotation of galaxies and the deduction that there must be a great deal of mass inside galaxies that cannot be seen. Cosmologists and physicists are not yet sure what dark matter is made of.

Our understanding of dark matter in the genome is much better because we know what it is made of (DNA) and we can isolate it and study its properties both directly and indirectly. One of the most powerful ways to study noncoding "dark" DNA is to hook a piece of it up to a gene that encodes a protein that is easily visualized, such as an enzyme that will make a colored reaction product, or a protein that will fluoresce in a beam of light. By inserting these engineered pieces of DNA back into a genome and then visualizing color patterns in a microscope, we can see what instructions, if any, are contained in a given piece of dark matter (a stripe here, a spot there, etc.). Most dark matter contains no instructions and is just space-filling "junk" accumulated over the course of evolution. In humans, only about 2 to 3 percent of our dark matter contains genetic switches that control how genes are used. I will focus all of this chapter on how genetic switches work to control animal development, and much of the rest of the book on how changes in these switches shape evolution.

I introduced the concept of genetic switches in chapter 3 by describing the genetic system for using lactose and an *E. coli* bacterium. Recall that in bacteria the synthesis of the enzymes for importing and breaking down lactose is controlled by a genetic switch. The switch is made up of a stretch of DNA sequence that lies just upstream of the genes that encodes these enzymes. When lactose is absent, the lac repressor binds to a specific DNA sequence in the switch and shuts off transcription; when lactose is present, the repressor falls off the switch, allowing the gene for breaking down lactose to be turned on.

In animals, genetic switches are a bit more elaborate. Generally, individual switches in animals are longer sequences of DNA, and they are bound by a larger number and greater variety of proteins. Some of these proteins activate transcription, some repress it. It is by "computing" the inputs of multiple proteins that switches transform complex sets of inputs into the simpler outputs we see as three-dimensional on/off patterns of gene expression, such as the stripes and spots in chapter 4. Importantly, one gene may be regulated by many separate switches such that the gene is used many times and in different places—for example, in the development of the heart, eyes, and fingers (figure 5.1).

The existence of switches expands our picture of how genes work.

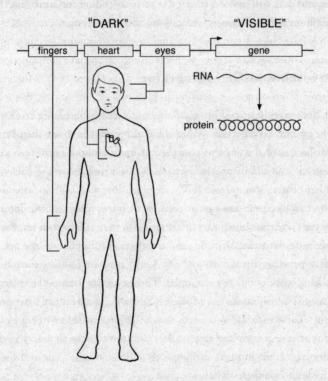

FIG. 5.1 **Genetic switches control where genes are used in body tissues.** This gene has switches that control its expression in the heart, eyes, and fingers. The presence of multiple switches, active in different developing body parts, is typical of tool kit genes. DRAWING BY LEANNE OLDS

Usually, when biologists talk about a gene, they mean specifically the stretch of DNA that is decoded into a protein, which in turn does work in cells. Switches are not decoded into anything—their function is regulatory in DNA. To carry out all of its normal functions, a gene depends on information coming from all of its switches. So a gene with three switches has four separable parts, one coding part and three regulatory parts (figure 5.1). Mutations in individual switches can cause some spectacular anatomical effects. I will continue with the typical usage of a "gene" as describing the protein-coding function, and I'll always make it clear when I am talking about switches.

Switches as GPS Integrators

We have seen that tool kit genes are activated in reference to three-dimensional coordinates within the embryo. But how are the spatial coordinates of the embryo conveyed as instructions to genes, to turn them on and off in precise patterns? The genetic switches act like global positioning system (GPS) devices. Just as a GPS locator in a boat, car, or plane gets a positional fix by integrating multiple inputs, switches integrate positional information in the embryo with respect to longitude, latitude, altitude, and depth, and then dictate the places where gene are turned on and off. I will explain and illustrate how switches work with a few examples. These examples should be thought of as just a few frames out of the whole movie of an animal's development. The entire show involves tens of thousands of switches being thrown in sequence and in parallel. We aren't going to worry about every frame; the important thing is to understand the logic and specificity built into these switches.

The general function of a switch is to transform existing patterns of gene activity into a new pattern of gene activity. One of the best examples to illustrate the working of a genetic switch is how a band or stripe of longitude is specified along the east-west axis in a fly embryo. Early in development, wide bands of 15–25 cells express specific tool kit proteins at different positions along the axis. Each individual tool kit protein binds to a specific DNA sequence, typically about 6–9 base pairs in length. The recognition of DNA sequences by tool kit proteins is similar to the way a specific key fits into a particular lock. In this case the lock is a particular DNA sequence. I will refer to these as "signature" sequences because they differ for each tool kit protein. Switches that control certain genes contain copies of these signature sequences and are occupied by the respective tool kit proteins in the nuclei of cells, at the longitudes and latitudes in the embryo where the tool kit proteins are present. In the example in figure 5.2, tool kit protein A is expressed from 20° to 60° W, protein B from 40° to 60° W, and protein C from 30°

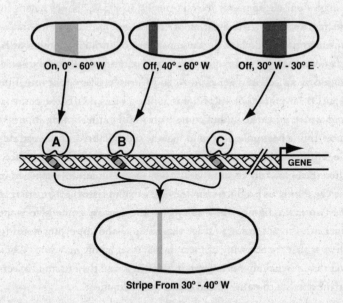

FIG. 5.2 **Switches integrate multiple inputs to draw a stripe of gene expression.** An activator (A) and two repressors (B and C) are expressed at different longitudes; the net output of the switch is a narrow stripe. DRAWING BY JOSH KLAISS

W to 30° E. Protein A is an activator while proteins B and C are repressors of gene X. In general, the rule will be that wherever repressors exist, they will cancel out the activators and the gene will be off. The switch for gene X contains sites for proteins A, B, and C. These sites will be occupied in different combinations along the axis.

In cells from 90° to 60° W, none of these proteins are on the switch and the gene is off; in cells from 60° to 40° W, both proteins A and B occupy the switch and the gene is off; in cells from 40° to 30° W, only protein A occupies the switch and the gene is on; and in cells from 0° to 30° E, only protein C occupies the switch and the gene is off. By "computing" three longitudinal inputs the switch allows the gene to be on only in a stripe just 10° wide, thus translating three broad patterns of gene expression into one narrow stripe. This stripe is positioned not by a single "on" cue that says "be on from 30° to 40° W," but by having its boundaries set by a combination of "off" inputs.

You might ask, where do these patterns of tool kit proteins A, B, and C come from? Good question. These patterns are themselves controlled by switches in genes A, B, and C, respectively, that integrate inputs from other tool kit proteins acting a bit earlier in the embryo. And where do those inputs come from? Still earlier-acting inputs. I know this is beginning to sound like the old chicken-and-the-egg riddle. Ultimately, the beginning of spatial information in the embryo often traces back to asymmetrically distributed molecules deposited in the egg during its production in the ovary that initiate the formation of the two main axes of the embryo (so the egg did come before the chicken). I'm not going to trace these steps—the important point to know is that the throwing of every switch is set up by preceding events, and that a switch, by turning on its gene in a new pattern, in turn sets up the next set of patterns and events in development.

Switches may integrate potentially any combination of longitude, latitude, altitude, and depth. An example of a switch that integrates inputs from different axes (figure 5.3) illustrates the actual mechanism of how limb positions are specified in the fly embryo. A switch in the

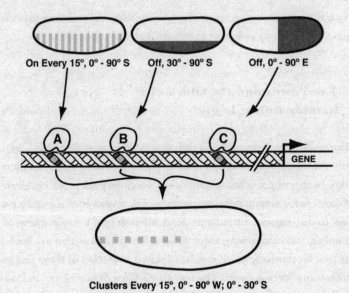

On Every 15°, 0° - 90° S   Off, 30° - 90° S   Off, 0° - 90° E

A   B   C

GENE

Clusters Every 15°, 0° - 90° W; 0° - 30° S

FIG. 5.3 Integration of longitude and latitude determines positions of small clusters of cells that will become limbs. DRAWING BY JOSH KLAISS

Distal-less limb-building gene integrates both longitude and latitude inputs to place several spots of *Dll* expression along the main body axis. These inputs are derived from several preexisting patterns of different types of tool kit proteins. One activator is distributed every 15° along the east-west axis within every segment, but only in the southern hemisphere (0°–90° S). Two different repressors are distributed from 30° to 90° S, and at all eastern longitudes, respectively. The integration of these three inputs produces a pattern of *Dll* expression in small clusters of cells at 90°, 75°, 60°, 45°, 30°, and 15° W, and at 0° to 30° S.

The physical integrity of switches is very important to normal development. If a switch is disrupted or broken by mutation, then the proper inputs are not integrated. Many of the spectacular mutants we've seen—flies with legs coming out of their head or humans with six fingers or toes—are due to broken switches that turn on tool kit genes in the wrong positions within the embryo or body part.

The Power and the Glory of Combinatorial Logic

The makeup of every switch is different. An average-size switch is usually several hundred base pairs of DNA long. Within this span there may be anywhere from a half dozen to twenty or more signature sequences for several different proteins. The response of a switch to a longitude, latitude, altitude, or depth input depends on the presence, number, and local arrangement of signature sequences that are bound by tool kit proteins, which may be deployed along any of these axes or within any specific tissue. The specific patterns drawn by any individual switch are determined by the specific sets of signature sequences encoded in the switch DNA.

In order to appreciate the information residing within a switch and the huge potential variety of switches, I need to provide a bit more detail about the nature of tool kit proteins and these signature sequences in switches. What follows is a brief explanation of the power of using the same tools in different combinations. The exact math isn't crucial, but understanding the power and efficiency of combinatorial logic is paramount.

The signature sequences recognized by tool kit proteins are short, usually about 6–9 base pairs in length, but can be longer. A lot of different signature sequences can fit into the span of one average-size switch. There are many different possible signature sequences. A 6-base-pair sequence has 4096 permutations of the four DNA bases A, C, G, and T (the math is $4^6 = 4096$), a 7-base-pair-sequence $4^7 = 16,384$ permutations, and an 8-base-pair sequence $4^8 = 65,536$ permutations. A given tool kit protein usually recognizes a family of closely related base sequences. There is some flexibility in the individual bases within a signature sequence but, even with this flexibility, tool kit proteins are highly selective in where they bind along DNA molecules. Different tool kit proteins generally recognize different signature sequences. Here is a very brief list of just a few tool kit proteins that bind to DNA and the signature sequences that they recognize:

Pax-6 (eyeless)	KKYMCGCWTSANTKMNY
Tinman	TCAAGTG
Ultrabithorax	TTAATKRCC
Dorsal	GGGWWWWCCM
Snail	CAGCAAGGTG

Where:

$$R = A \text{ or } G$$
$$Y = C \text{ or } T$$
$$K = G \text{ or } T$$
$$M = A \text{ or } C$$
$$S = C \text{ or } G$$
$$W = A \text{ or } T$$
$$N = A, C, G, \text{ or } T$$

The whole tool kit of an animal contains several hundred or so different DNA-binding proteins, most with different signature preferences. There are an astronomical number of potential combinations of signature sequences in switches. If we assume a tool kit of 500 DNA-binding proteins in an animal, there are 500 x 500 = 250,000 different pairs of combinations of sequences and tool kit proteins. There are 500 x 500 x 500 = 12,500,000 different three-way combinations and over 6 billion different four-way combinations. These calculations illustrate the power of combinatorial logic of the tool kit and genetic switches. The great variety of switches is a product of using the same signature sequences and tool kit proteins in myriad combinations. One could imagine the alternative would be to have a larger number of tool kit proteins, but it is far more efficient to use 500 proteins in combinations than to encode 250,000 different proteins (which is about ten times more proteins than are encoded in our entire genome).

Allow me one brief aside on the power of combinatorial logic in biology. We have seen this power before in an entirely different context. Our immune systems cope with the enormous diversity of the potential pathogens that live within and around us by making antibody proteins that bind to the proteins, sugars, and fats of these foreign invaders. We have the capacity to make millions of different kinds of antibody proteins. This huge capacity is generated by combining a modest number (a few hundred) of antibody gene regions and antibody chains in different ways, not by encoding millions of different individual antibody genes.

The versatility of switches and of combinatorial logic is very clearly illustrated by experimenting with the DNA sequence content of switches. Simply adding signature sequences to or subtracting them from switches and observing how the patterns they draw change vividly demonstrates the flexibility and power of switches. Mike Levine and his colleagues at the University of California–Berkeley have been leaders in exploring the combinatorial logic of making stripes along both axes in the fly embryo and their work has revealed simple but elegant mechanisms for making patterns.

The basic logic of stripe-making in the early fly embryo was shown in figure 5.2 for a longitudinal stripe. The same general idea of drawing stripes also operates along latitudes. The exact position of a stripe depends upon the strength of the inputs into switches. One way to increase the strength of an input is to add more copies of a signature sequence to a switch. For example, a latitudinal (horizontal) stripe of gene expression that occurs along the southernmost extent of the fly embryo is activated by a tool kit protein whose concentration is graded from south to north. The switch normally contains two signature sequences for this protein. When two copies of a signature sequence for another protein are added to this switch, the stripe expands to more

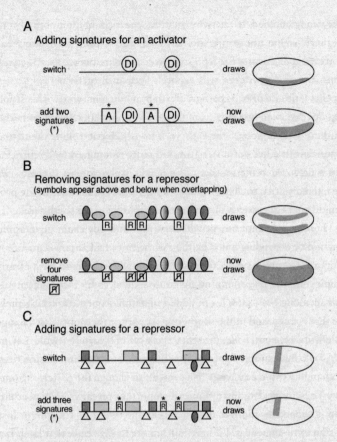

FIG. 5.4   A. Adding sites for a repressor removes part of the pattern of gene expression. B. Removing sites for a repressor expands the pattern of gene expression. C. Adding sites for a repressor removes part of the pattern of gene expression. DRAWING BY JOSH KLAISS

than twice its original width, covering more of the southern hemisphere of the embryo (figure 5.4A).

Alternatively, input can be weakened by reducing the number of signature sequences present in a switch, or it can be eliminated altogether. If these signature sequences are for activators, then the switch can be completely crippled. If the two signature sequences in the switch for the southern stripe above are altered, then this switch is inactivated. However, if signatures for repressor proteins are removed, then the patterns drawn by switches will expand. Section B of figure 5.4 shows another latitudinal stripe pattern drawn by a different switch. This stripe is about 20° wide, extends from 40° to 60° S, and is excluded from the southernmost region of the embryo. The switch controlling this stripe contains four copies of a signature sequence recognized by a repressor that is deployed in the southernmost region of the embryo. If these sites are altered so that the repressors no longer binds to the switch, the pattern drawn by the switch then extends completely to the south pole.

These simple experiments demonstrate how the exact geographic position of a stripe is tuned by the assortment of signature sequence it contains. Getting a switch to draw patterns with respect to both axes is simply a matter of containing signature sequences for tool kit proteins that act along both axes. If one adds a signature sequence for this repressor that is expressed in the southernmost part of the embryo to a longitude stripe element, *voilá*, the stripe is cut off in the south (figure 5.4C).

All of these simple experiments illustrate how adding, subtracting, or changing just a few bases in a switch can change the patterning output. These nifty demonstrations are important previews of how evolution is shaped by changes in switches by the evolutionary gain or loss of signature sequences. I'll have much more to say about that later, but it is well worth beginning to think about the possibilities as we delve into the world of genetic switches.

Stripe by Stripe, Bone by Bone: The Whole Is the Sum of Many Parts

The genes expressed in stripes in the early fly embryo were some of the first to have their switches examined. One of the most surprising discoveries made in taking these genetic switches apart was that individual stripes of multistripe patterns are encoded by separate switches. For example, even though the seven stripes of some tool kit patterns appear very similar and evenly spaced, each stripe is drawn by a different switch that integrates different combinations of longitudinal inputs. This seemed at first like an awful lot of machinery for making just one pattern. But this stripe-by-stripe construction of striped patterns in the fly embryo was the first clue to the general rule that the whole expression pattern of any tool kit gene is actually the sum of many parts, with individual parts controlled by individual switches.

The revelation of how these stripe-making switches work clarified a long-standing question in the study of pattern formation in biological structures. For several decades, mathematicians and computer scientists were drawn to the periodic patterns of body segmentation, zebra stripes, and seashell markings. Heavily influenced by a 1952 paper by the genius Alan Turing (a founder of computer science who helped crack the German Enigma code in World War II), "The Chemical Basis of Morphogenesis," many theoreticians sought to explain how periodic patterns could be organized across entire large structures. While the math and models are beautiful, none of this theory has been borne out by the discoveries of the last twenty years. The mathematicians never envisioned that modular genetic switches held the key to pattern formation, or that the periodic patterns we see are actually the composite of numerous individual elements.

A gene not only may have multiple switches for different subpatterns of expression at a given time, but will frequently have different switches that control entirely different patterns in different tissues and at different stages in development. Tool kit genes are rarely, if ever, devoted to a single developmental operation. Rather, these tools are used and reused again and again in development in different contexts to shape the growing embryo. Switches endow individual tool kit genes with great versatility. Virtually every tool kit gene is controlled by multiple switches. Ten switches or more is not uncommon, and we don't know what the upper limit, if any, may be.

The building of bodies and body parts is accomplished by the sum of operations governed by individual switches. The large and complex skeletal anatomy of vertebrates is actually encoded and constructed, bone by bone, by arrays of switches nested around a host of tool kit genes. One family of important tool kit proteins for skeletal development, the Bone Morphogenetic Proteins (BMPs), is so called because they have the property of promoting cartilage and bone formation. The regulation of one member of this family, the *BMP5* gene, vividly illustrates how parts of anatomy are encoded in pieces through separate switches.

All around the *BMP5* gene, there are switches. There are separate switches for BMP5 expression in ribs, limbs, fingertips, the outer ear, the inner ear, vertebrae, thyroid cartilage, nasal sinuses, the sternum, and more (figure 5.5). The same protein is being produced in all of those different patterns, places, and times—the specificity of each operation and the complexity of the overall pattern are entirely due to the array of switches. The existence of separate switches for each of these parts illustrates the fine-tuning control that is available for the construction and shaping of every body part.

A Cornucopia of Switches

The spectacular diversity and exquisite geographic specificity of switches derive from the use of combinatorial logic. Because the combination of inputs determines the output of a switch, and the potential combinations of inputs increase exponentially with each additional input, the potential outputs of switches are virtually endless. Imagine the possibilities of combining bands, stripes, lines, spots, dots, and patches of activators and repressors and the ability to draw these in any place, in any tissue, and in any combination. All kinds of patterns are possible, and switches that draw an enormous variety of patterns have been found in individual animal genomes. For any coordinate or

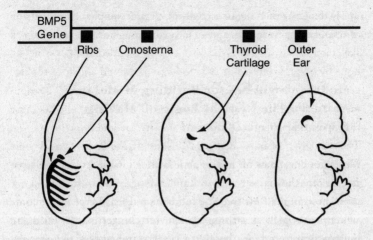

FIG. 5.5 Individual switches control the expression of the *BMP5* gene in different parts of the developing mouse embryo. ADAPTED FROM DAVID KINGSLEY, HOWARD HUGHES MEDICAL INSTITUTE AND STANFORD UNIVERSITY; DRAWING BY JOSH KLAISS

sets of coordinates, switches can and do draw just about any geometric pattern of gene expression.

While it is true that the number of potential combinations of inputs and signature sequences is enormous, the actual set of switches in any animal is finite. And not every switch is entirely different. In order to coordinate development, especially the making of particular cell types that have dedicated functions, switches in different genes often share one or more inputs and signature sequences in common. For example, in order to function as muscle cells, a set of proteins must be produced that enable the cells to contract, to rapidly utilize energy sources, and to efficiently remove waste during muscle activity. The genes encoding these proteins are activated in the muscle cell by switches with common signature sequences recognized by the same tool kit protein. The same is true in other specific cell types—neurons, photoreceptor cells in the eye, pancreatic cells, the pituitary, etc. Organ functions commonly depend upon one or a few tool kit proteins that throw sets of switches belonging to many genes located throughout the genome.

Modular Switches for Building Modular Animals: The Critical Logic of Making Repeated Parts Different

With a good feel now for how genetic switches work, let's turn to how they fit into the major trends of animal design, and to start thinking about how animals evolve. The fundamental feature of large, complex animals such as arthropods and vertebrates is their modular construction from repeating parts. Understanding how switches are used to make repeating parts into different forms with different functions is central to understanding the making and evolution of our favorite animals.

We saw in the last chapter that the expression of different *Hox* genes occurs in the different segments and appendages of arthropods, and the different rhombomeres and somites of vertebrates. The pattern and function of each repeated part depend upon the unique *Hox* gene or combination of *Hox* genes acting in each segment, appendage, somite, or rhombomere. The establishment of these *Hox* "zones" and their subsequent action in sculpting the different forms of repeated parts is the fundamental genetic logic upon which the modular forms of large, bilateral animals are built.

This genetic logic relies on genetic switches at two levels. One set of switches belongs to the *Hox* genes themselves. These switches activate each *Hox* gene in different zones that will become different modules of the animal. Another set of switches contain signature sequences that are recognized by Hox proteins and that control how other genes are expressed in different modules.

In both arthropods and vertebrates, the *Hox* genes are deployed in zones along the main body axis. The distinct zones of each *Hox* gene's expression domain are governed by genetic switches, and separate switches control *Hox* gene patterns in different tissues such as the hindbrain, neural tube, somites, and limb buds in vertebrates and the epidermis and nerve cord in arthropods. Because of the logic of these switches, the cells that belong to one module express different Hox proteins or combinations of Hox proteins than those in adjacent modules. The different forms of each module—brain rhombomere or somite, arthropod segment or appendage—are sculpted by Hox proteins acting on other genes.

The general logic of how Hox proteins sculpt the different morphologies of repeated parts is most easily illustrated in insects. Along the main body axis, most segments are patterned differently and bear different structures. For example, the first thoracic segment bears no wings, the second thoracic segment bears the large forewings, and the third thoracic segment bears the smaller hindwings used for balance. No Hox protein is expressed in forewing cells but all hindwing cells express Ubx (because a set of switches in the *Ubx* gene activate it in the third thoracic segment and hindwing). The difference in appearance between the hindwing and forewing is due to Ubx action.

Ubx sculpts the form of the hindwing by acting on the switches of genes that pattern the wings. It turns off genes that promote the formation of forewing features (veins and other structures) and turns on genes that promote hindwing features. The switches in these genes must integrate multiple inputs (and contain signature sequences for each). If we take a snapshot of a handful of switches and gene activity and contrast their states in the forewing and hindwing, the basic logic that we find is that Ubx acts on a subset of switches to shape the hindwing to be different from the forewing (figure 5.6).

The same logic applies to making different rhombomeres, different limb types in arthropods, vertebrae, and ribs. The different final forms of these serially reiterated structures are sculpted by Hox proteins that

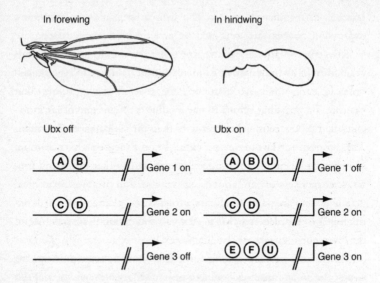

In forewing

In hindwing

Ubx off

Ubx on

(A)(B) // Gene 1 on

(A)(B)(U) // Gene 1 off

(C)(D) // Gene 2 on

(C)(D) // Gene 2 on

// Gene 3 off

(E)(F)(U) // Gene 3 on

FIG. 5.6 **Alternative states of gene expression in the forewing and hindwing are controlled by a Hox protein.** The solid lines represent switches, the letters different regulatory proteins (U stands for Ubx). The different forms of the two wings result from different sets of genes being active in each. DRAWING BY JOSH KLAISS

determine which subsets of limb-, rhombomere-, vertebrae-, or rib-patterning genes are active at each position along the body axis.

The "Wiring" of the Embryo: Switches, Circuits, and Networks

I have illustrated the way genetic switches work by focusing on one switch of a given gene, the various switches of one gene, or the assortment of switches controlled by a common protein. But every switch or protein I have described and each pattern I have shown is just a still photo—adding up to relatively few frames in the whole course of an animal's development. The entire story of making an animal has many, many more frames—it is one hell of a movie with nonstop action.

The forms of animals and their body parts are never the result of the action of a single switch or protein. Body parts, tissues, and cell types are the products of large numbers of switches and proteins that organize patterns in time and space, and of proteins and other molecules that endow cells and tissues with their physiological and mechanical properties. The developmental steps executed by individual switches and proteins are connected to those of other genes and proteins. Larger sets of interconnected switches and proteins form local "circuits" that are part of still larger "networks" that govern the development of complex structures. Animal architecture is a product of genetic regulatory network architecture.

This circuit and network wiring or logic can be illustrated with the same sort of diagrams used for electrical circuits or logic problems. Each

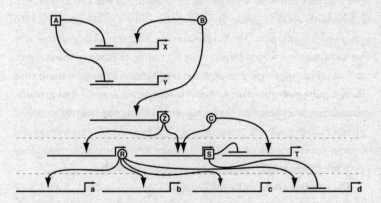

FIG. 5.7 **A genetic wiring diagram of regulatory logic.** Activators (circled letters) and repressors (squared letters) act on switches (solid lines). Arrows indicate activation effects, lines ending with a perpendicular line denote repression. Multiple tiers of activators and repressors are usually involved in building and patterning any structure. DRAWING BY JOSH KLAISS

switch is a decision point, one node in the genetic circuitry. Figure 5.7 shows a set of interconnected circuits that involve a small number of activators, repressors, switches, and genes. This again is a model of just a few parts of a much bigger picture. My guess is that I would need at least one thousand pages to write out the logic of making a fly, and several thousand pages to write out the making of a human. Regulatory networks in vertebrates are more numerous (we have three times as many cell types as flies or other invertebrates), but not really any more complicated.

Switches and Solving the Tool Kit Paradox

Biologists are still coming to grips with the profound importance of genetic switches. For several decades, we have been able to read the genetic code and see exactly how and where protein sequences are encoded in DNA. The common view from this protein-centric perspective was that genes were bodies of information in the vast expanse of DNA, with all that space around and between genes being largely empty of information. The belief was also widespread that differences between animals would largely be a matter of changes in the number and sequence of genes. But now we are beginning to understand that there can be many genetic switches surrounding a gene. And genome sequencing has shown us that mice and humans have nearly identical numbers and kinds of genes (about 25,000 each). So, given that the coding sequences are so similar, it is time to explore the surrounding switches to understand their roles in evolution.

The glimpses here into the logic and great potential diversity of genetic switches prepare us to start thinking about their contribution to the evolution of animal diversity. The great paradox raised by the discovery of similar sets of tool kit genes in disparate animals is how the same genes can be used to build such different forms. The discovery of arrays of switches that enable individual tool kit genes to be used again and

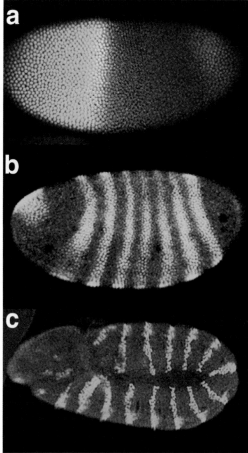

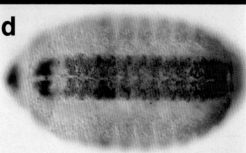

4a. The western and middle regions of the early fly embryo are marked by the expression of two tool kit proteins (in green and red; overlap appears yellow). Each filled circle is an individual cell nucleus.

4b. The fly embryo is then subdivided into double-segment intervals by a set of tool kit proteins expressed in one stripe per every two future segments.

4c. The embryo is then subdivided into segmental intervals; this tool kit protein marks the back part of every future segment. PLATES A–C JIM LANGELAND AND STEVE PADDOCK

4d. Zones of Hox protein expression are established at different longitudes. Four Hox proteins are revealed here (in four colors). PHOTO COURTESY OF NIPAM PATEL, UNIVERSITY OF CALIFORNIA–BERKELEY

4e. The latitudes of the early embryo are subdivided by tool kit genes expressed in the southernmost (top), equatorial (middle), and northern (bottom) regions of the embryo. COURTESY OF MICHAEL LEVINE, UNIVERSITY OF CALIFORNIA– BERKELEY

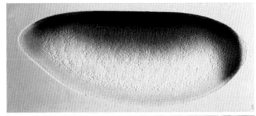

4f. The future positions of appendages are marked by tool kit protein expression at specific intersections of longitude and latitude. The position of the future forewing (w) and hindwing (h) and legs (l) are marked by two different proteins. SCOTT WEATHERBEE

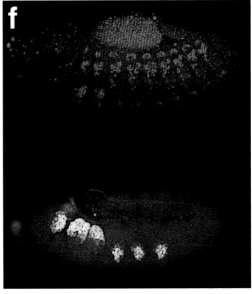

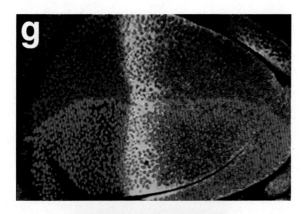

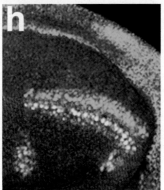

4g. The growing wing is subdivided into its future upper (red) and lower (magenta) surfaces and back (left) and front (yellow and right) sections by different tool kit proteins. Every filled circle is the nucleus of a wing cell. JIM WILLIAMS AND STEVE PADDOCK

4h. At particular coordinates in the wing, tool kit proteins that promote the formation of particular structures are turned on. Here, the tool kit protein shown in yellow and green is promoting sensory bristle formation along the future edge of the wing (where the red zone ends). SETH BLAIR, UNIVERSITY OF WISCONSIN

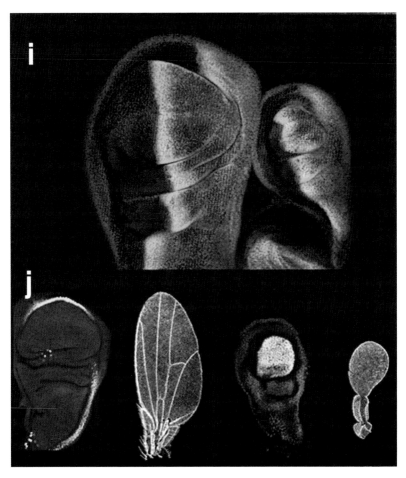

4i. The forewing and hindwings, although different in size, are subdivided by the same tool kit proteins (compare purple, red, and green/yellow patterns between the two wings). JIM WILLIAMS AND STEVE PADDOCK

4j. A Hox protein distinguishes the hindwing (far right) from the forewing (second from left). The large disc at left will make the wing; none of the future wing cells express Ultrabithorax (yellow). All cells that will make the hindwing (intense yellow in disc at second from right) express the Hox protein. SCOTT WEATHERBEE

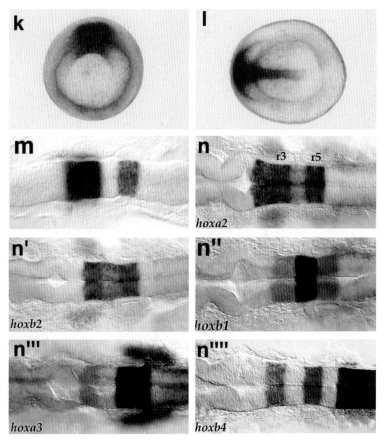

4k. The tool kit protein Chordin is made in the organizer cells of the frog embryo. COURTESY OF EDDY DE ROBERTIS, UCLA

4l. The Frzb tool kit protein is made in cells toward the head region of the frog embryo. COURTESY OF EDDY DE ROBERTIS, UCLA

4m. Tool kit genes mark the future subdivisions of the vertebrate hindbrain. Three genes are revealed here (in blue, black, and orange) whose expression mark rhombomeres r2, r3, and r5. COURTESY OF CECILIA MOENS, HOWARD HUGHES MEDICAL INSTITUTE, FRED HUTCHINSON CANCER CENTER, SEATTLE

4n. Zones of *Hox* expression in the vertebrate hindbrain. In each panel, r3 and r5 are marked in pink/orange by the *Krox 20* tool-kit gene. Five different *Hox* patterns are shown in purple in the hindbrain and have different boundaries from r2 (*Hoxa2*) to r7 (*Hoxb4*). IMAGES COURTESY OF CECILIA MOENS, HOWARD HUGHES MEDICAL INSTITUTE, FRED HUTCHINSON CANCER CENTER, SEATTLE

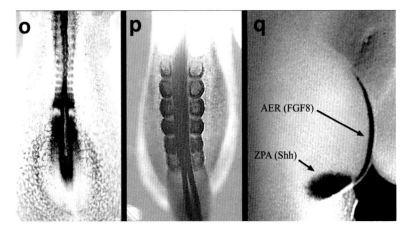

4o. The making of somites is marked by the expression of a tool kit gene in each developing somite, including those yet to be shaped. COURTESY OF OLIVIER POURQUIE, STOWERS INSTITUTE, KANSAS CITY, MISSOURI

4p. *Hox* zones are established at particular somites along the main trunk of the body. COURTESY OF OLIVIER POURQUIE, STOWERS INSTITUTE, KANSAS CITY, MISSOURI; REPRINTED FROM *CELL* 106 (2001): 219–32, BY PERMISSION OF ELSEVIER

4q. Expression of a tool kit gene that marks the position of the developing wing and leg buds. COURTESY OF JOHN FALLON, UNIVERSITY OF WISCONSIN

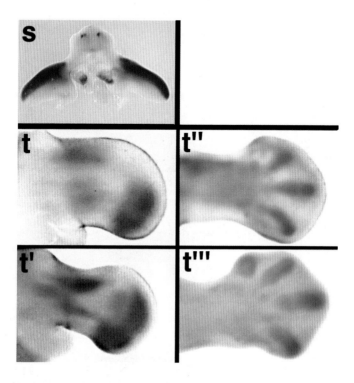

4r. Two key regions of the chick limb bud are marked by expression of tool kit genes. The *Sonic hedgehog* gene is expressed in the ZPA, the *FGF8* gene in the ridge of the limb bud. COURTESY OF CLIFF TABIN, HARVARD MEDICAL SCHOOL, CAMBRIDGE, MASSACHUSETTS

4s. The *Lmx* tool kit gene marks the upper halves of the future limb. Two limb buds are shown; note the purple stain in the upper half of each. COURTESY OF CLIFF TABIN, HARVARD MEDICAL SCHOOL, CAMBRIDGE, MASSACHUSETTS

4t. Progression of limb development and cartilage formation revealed by expression of a tool kit gene. In the first panel, the expression of the *Sox9* gene marks the formation of the upper limb at the base of the bud. The remaining panels reveal the progressive definition of the forearm, hand, and digits. COURTESY OF DR. JUAN HURLE, UNIVERSIDAD DE CANTABRIA, SANTANDER, SPAIN; REPRINTED FROM *DEVELOPMENTAL BIOLOGY* 257 (2003): 292–301, BY PERMISSION OF ELSEVIER

4u. The *GDF5* tool kit gene marks the future joints in the digits. COURTESY OF DR. JUAN HURLE, UNIVERSIDAD DE CANTABRIA, SANTANDER, SPAIN; REPRINTED FROM *DEVELOPMENTAL BIOLOGY* 257 (2003): 292–301, BY PERMISSION OF ELSEVIER

4v. The *scleraxis* tool kit gene marks the future tendons of the limb and digits. COURTESY OF CLIFF TABIN, HARVARD MEDICAL SCHOOL, CAMBRIDGE, MASSACHUSETTS

4w. The *BMP4* tool kit gene marks the tissue between the digits that will die. COURTESY OF CLIFF TABIN, HARVARD MEDICAL SCHOOL, CAMBRIDGE, MASSACHUSETTS

4x. The *patched* tool kit gene marks the position of developing feather buds along the back of the developing chicken. COURTESY OF CLIFF TABIN, HARVARD MEDICAL SCHOOL, CAMBRIDGE, MASSACHUSETTS

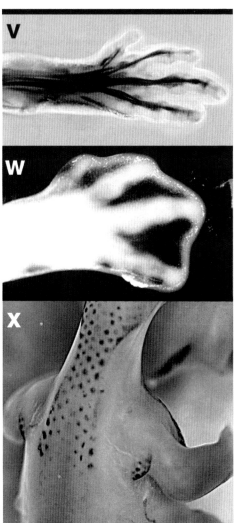

again in one animal, and to be used in slightly or dramatically different ways in serially repeated structures, is key to solving this paradox.

It is a small leap from understanding how switches control development to anticipating how they have shaped evolution. *Switches enable the same tool kit genes to be used differently in different animals.* Because individual switches are independent information-processing units, evolutionary changes in one switch of a tool kit gene or in a switch controlled by a tool kit protein can alter the development of one structure or pattern without altering other structures or patterns. This is the key to the evolution of modular bodies and body parts—how we, for example, can evolve an opposable thumb, or flies can evolve a special hindwing. Many of the evolutionary mysteries I will now explore in the second part of the book, from the great burst of diversity in animal forms that marks the Cambrian Explosion to the wonderful variety of butterflies or mammals living today, were shaped by evolutionary changes in genetic switches.

Part II

Fossils, Genes, and the Making of Animal Diversity

A DECADE OR SO AGO, molecular biologists, "indoor" laboratory-oriented folks like myself who played with DNA, and paleontologists, "outdoor" field scientists who traveled to exotic locales and extracted ancient treasure from rocks, were complete strangers. With almost nothing in common, we never met, let alone dated. We were trained differently, usually worked in entirely different university departments, and published in different scientific journals.

All of this has changed.

Now paleontologists talk of *Hox* genes and molecular biologists even dare use words like "Cambrian" in a sentence!

In the second part of this book, I will tell the very happy story of the union of embryology with evolutionary biology in solving the mysteries of the evolution of animal forms. The trigger for this union was, in large part, the powerful technologies of molecular biology that have provided entirely new means of looking at animal development and history. Knowledge of living animal genomes and the development of embryos allows us to look at animal history as depicted in the fossil record with a new perspective, and to make new insights into not just what happened but *how*—to perceive the inner workings of the making of animal diversity. One of the founding tenets of modern geology was that "the present is the key to the past"—the idea that processes we can observe now operated in and explain the past. This basic idea is also one of the fundamental principles of the new science of Evo Devo.

The first part of this book has set the stage by illustrating four critical ideas about animal development—the modularity of animal architecture, the genetic tool kit for building animals, the geography of the embryo, and the genetic switches that determine the coordinates of tool kit gene action in the embryo.

In the second half of the book, the central idea is that animal forms evolve through changes in embryo geography. We will learn the specifics of how geography and form evolve by changing the way the tool kit genes are used. Evolution of form is very much a matter of teaching very old genes new tricks!

In the course of the next chapters, we will learn about the power of Evo Devo to peer into the distant past to help us draw pictures of long extinct animal ancestors and to shed light on some of the most dramatic episodes in animal history. We will examine evolution from the deepest roots of the animal kingdom, which first emerged in ancient seas more than 500 million years ago, to the origins of new structures that allowed new types of animals to make a living on the land and in the air, to the most recent twigs on the animal tree that make up the spectacular diversity of today's animals. This builds the framework for the exploration of how we humans evolved from a small-brained, quadripedal hominid ancestor.

The stories I will tell create a vivid new picture of the evolutionary process. The impact of Evo Devo comes from both its novelty and the unprecedented quality of evidence it provides. Some of this new evidence conclusively settles long-running debates in evolutionary biology, some raises entirely new ideas, while other discoveries have revealed one of the "Holy Grails" of evolutionary biology—the precise genetic changes responsible for evolution in particular species.

Because embryology, through Evo Devo, now holds a costarring role in a fully integrated evolutionary synthesis, it is also time to change the textbooks to reflect this revolution. Indeed, I believe that Evo Devo's methods of illustrating how animal forms evolve offers a much more powerful explanatory vision than the abstract extrapolations of the era of the Modern Synthesis. To the classic evolutionary tales of natural selection in Galapagos finches and peppered moths, Evo Devo now adds deep insights from lobsters and shrimp, spiders and snakes, spotted butterflies, pocket mice, and jaguars. And it illustrates as never before how Darwin's "endless forms" have been and are being made.

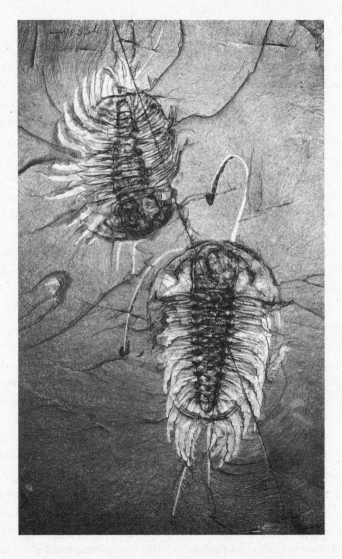

Olenoides serratus trilobites of the Burgess Shale. PHOTO BY CHIP CLARK, BY PERMISSION OF SMITHSONIAN INSTITUTION

6

The Big Bang of Animal Evolution

It seems that Nature has taken pleasure in varying the same mechanism in an infinity of different ways. . . . She abandons one class of
production only after having multiplied the individuals of it in all possible forms.

—Denis Diderot, *Pensées sur l'Interpretation
de la Nature* (1753)

JUST INSIDE THE ENTRANCE to the hall of fossil exhibits at the Smithsonian's National Museum of Natural History in Washington, D.C., there is a set of institutional green, ordinary glass cabinets. Most visitors stride right by, drawn by the gallery of dinosaurs and other dramatic behemoths just beyond. But inside the nondescript cases, in small square rock blocks, lie some of the most extraordinary and important animal fossils ever discovered.

These are the fossils of the Burgess Shale. First uncovered by Charles Walcott of the Smithsonian on an expedition to British Columbia in 1909, the fossils date from the Middle Cambrian, about 505 million years ago. The weird and wonderful forms recorded in dark gray-black shale have long captivated paleontologists. They are among the earliest complex animals, complete with antennae, limbs, tails, and eyes, and they include representatives of many modern groups, including arthropods, annelids, chordates, and molluscs. And, they appear to have burst upon the scene in a relatively brief interval of about 15 to 20 million years, prior to which there are only scant examples of animal life in the fossil record. This episode of the geologically rapid appearance of complex forms is recorded in rocks 525 to 505 million years old found around the world—the so-called Cambrian Explosion—the Big Bang of animal evolution.

These fossil animals and the phenomenon of the Cambrian Explosion were first brought to broader public attention by the late Stephen Jay Gould in his superb book *Wonderful Life*. One of the first challenges presented by the Cambrian animals was to figure out to what group individual fossils belonged. Their "weird" anatomy (from a modern viewpoint) has led to many different and changing opinions about whether a given fossil is a mollusc or a worm, an arthropod or not, or something altogether different from anything we know today.

The relationships of these animals to modern groups is but one of many mysteries surrounding the Cambrian Explosion. The others would include: What ignited the Explosion? Why did large and complex animals first appear at this time? Why did these particular forms succeed? Many ideas have been put forth as to causes of the Explosion. Some theories focus on extrinsic explanations, such as a change in global climate. Other theories focus on intrinsic causes, such as the invention of body-building genes. Like many events in the deep past, it has been easier to conjure up ideas than to test them. With respect to genetic theories, how does one ask questions about the genes of animals that have been dead for more than 500 million years? Indeed, the Cambrian fossils aren't animals but just the impressions of animals squished under tremendous geological forces. But the dramatic progress in embryology has enabled us to figure out what roles genes played in the ignition and expansion of the Cambrian Explosion. The new power of Evo Devo is, in one sense, able to make long dead forms come alive.

The surprising message from Evo Devo is that all of the genes for building large, complex animal bodies long predated the appearance of those bodies in the Cambrian Explosion. The genetic potential was in place for at least 50 million years, and probably a fair bit longer, before large, complex forms emerged. This means that while the genetic tool kit was not evolving, the rapid appearance of and changes in body forms tell us that animal development was evolving a great deal.

The main story of many groups in the Cambrian is that of evolving different numbers and kinds of repeated body parts. This dramatic illustration of Williston's Law is explained by changes in embryo geography. Shifts in the coordinates of tool kit genes, particularly where the *Hox* genes are expressed in the embryo, are responsible for the making of different body forms. These shifts are brought about through genetic switches—the evolution of switches drove the Cambrian, and the subsequent evolution of major classes of animals that arose later.

The question of how different forms evolved will be front and center in this chapter. But to set the scene, we need to know what actually happened in animal history—what came before the Explosion, what transpired during it, and what followed after. I'll start with the issue of what forms existed prior to the Cambrian. Despite the scarcity of the earlier fossil record, Evo Devo allows us to peer even deeper into animal history, before the Cambrian, to ponder the complexity and forms of the ancestors of the Cambrian animals—especially the mysterious last common ancestor of all bilateral animals, including ourselves.

Building a Mystery: Animals Before the Big Bang

The Earth is about 4.5 billion years old. Perhaps as early as 3.5 billion years ago, life began to evolve but for the first 3 billion years organisms were generally small (on the order of millimeters or smaller) and of simple construction. Several kingdoms preceded our own—bacteria, archaea, protists, and fungi (land plants came later than animals although their forerunners, green algae, predate the animals). Late in the Precambrian, about 600 to 570 million years ago, the size and shape of life began to expand and various centimeter-size forms appear in the so-called Ediacaran fauna (named for the hills in South Australia where representatives of these forms were first found). This enigmatic group has stumped paleontologists for decades. Harvard biologist Andy Knoll likens them to a paleontological Rorschach test. The tube-shaped, frond-

FIG. 6.1 **Ediacaran forms.** *Dickinsonia costatala* and *Spriggina flounders* from the Ediacara Hills of South Australia are of uncertain relationship to modern or Cambrian forms. PHOTOS COURTESY OF DR. JIM GEHLING, SOUTH AUSTRALIAN MUSEUM; USED BY PERMISSION

like, and radially symmetrical forms of Ediacaran fossils have been asserted to be everything from artifacts of sedimentation, to an extinct experiment in multicellular life, to animal ancestors, as well as representatives of living groups; but their connection to modern animals remains uncertain and controversial (figure 6.1). Whatever the Ediacaran oddballs were, during this time the ancestors of the animals of the Cambrian must have existed. We don't know what they looked like but new insights from Evo Devo allow us to imagine what to look for.

To think about our ancestors we have to make some inferences based upon the structure of the animal evolutionary tree (figure 6.2). Biologists are concerned with the position of groups on evolutionary trees because knowledge of relationships allows us to deduce when features evolved, and in which groups. Insects and vertebrates are representatives of the two different main branches of the animal tree. These two branches are defined and distinguished by a fundamental difference between their embryos in where the mouth forms with respect to the initial blastopore opening. Those animals in which the mouth forms at a site separate from the blastopore opening in the embryo are called *deuterostomes* and include ourselves, all other vertebrates, echinoderms (e.g, sea urchins, sand dollars), and a few other groups. Those animals in which the mouth develops from the blastopore are called *protostomes* and include flies, other arthropods, annelids, molluscs, and a host of other groups. The trunk of the animal tree includes sponges, cnidarians (jellyfish, corals, sea anemones), and comb jellies that split off before protostomes and deuterostomes. (These "trunk" animals are important in life's history, and in the oceans today, but I'm not spending much time on them in this book; I'm more concerned with the two higher branches.)

The first appearance of clearly recognizable members of many protostome and deuterostome groups is in the Cambrian period. Because these branches of the animal tree are clearly distinguishable at this time, we deduce that the common ancestors of various groups must predate the Cambrian by some chunk of time. This is an inference

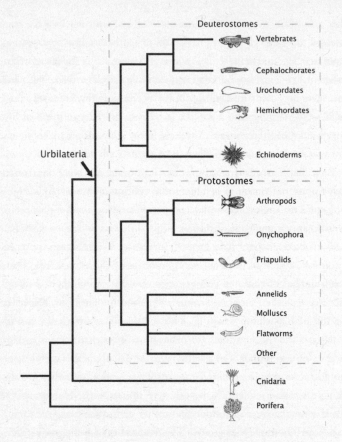

FIG. 6.2 **The animal evolutionary tree.** The main groups of bilateral animals are deuterostomes and protostomes, whose last common ancestor is referred to as Urbilateria. Cnidarians (sea anemones, corals) and Poriferans (sponges) split off before bilaterians. DRAWING BY JOSH KLAISS

because the fossil record of protostomes and deuterostomes before the Cambrian is very scanty. In fact, only one body fossil from the Precambrian, an animal called *Kimberella*, dated at 555 million years old, has been suggested to represent a protostome.

So, where then are those ancestors? Fossil preservation conditions were adequate to preserve animals such as jellyfish, corals, and sponges, as well as the Ediacaran fauna. It does not appear that scarcity is a fault of the fossil record. Because large fossils of some organisms do exist, one explanation for the lack of protostome and deuterostome fossils is that these animals were small (maybe less than one centimeter in size) and of delicate construction. Another possibility is that among the variety of weird Ediacaran fossils are protostomes or deuterostomes, but we just don't recognize them because they didn't yet have the features of more modern animals. Without confirmed body fossils, paleontology is reluctant to conjure up more than a vague image of a featureless, wormlike creature for the last common ancestor of protostomes and deuterostomes.

If we can't say much for certain from the fossil record, what can we say about animal ancestors based on other kinds of evidence? We can make inferences based on what is shared among descendants. This is the critical logic used in Evo Devo to peer into the distant past. The basic premise is that whatever is shared by two or more groups is likely to have existed in their last common ancestor (represented by the fork at the base of two branches in a tree). We can then apply this logic to our knowledge of development and genes in two or more groups to make inferences about features in common ancestors. One feature we can assert is that the last common ancestor of protostomes and deuterostomes was bilaterally symmetrical. All members of both groups have, at least at some stage in their lives, a bilateral organization (in sea urchins and other echinoderms, larvae are bilaterally organized, even though the adults show various radial symmetries). This organization opened up new means of locomotion and more complex lifestyles. But we can now go further than this: based on the content and similar roles of the shared genetic tool kit in protostomes and deuterostomes, we can confidently add that the common ancestor of bilaterians (an animal that Eddy De Robertis at UCLA has dubbed Urbilateria, meaning primitive bilaterian) had a tool kit of at least six or seven *Hox* genes, *Pax-6*, *Distal-less*, *tinman*, and a few hundred more body-building genes.

It is intriguing to ponder just what so many genes were doing in Urbilateria. Was this really a featureless wormlike animal? What might the possession of so many genes signify in terms of anatomical and behavioral complexity?

One way to account for the similar roles in different animals of genes in the tool kit is to posit some level of anatomical complexity in Urbilateria that was governed by these genes. That level is somewhat open to different interpretations but we can build up a picture of Urbilateria based on some reasonable inferences. For example, could Urbilateria have had eyes? Well, probably not the large pronounced eyes like those we find on trilobites later in the Cambrian. Something that had large, complex eyes would probably have turned up by now in the fossil record. But, because the role of *Pax-6* and other genes involved in eye development is shared in both major branches of bilaterians, we can deduce that Urbilateria probably had at least some kind of eyespot or light-sensing organ made up of photosensitive cells arranged in some geometry.

Using similar logic, we can ask if Urbilateria had limbs. Paleontologists can detect the traces of the meanderings of animals in fossil sediments yet they do not really become substantial until the Cambrian, so full-fledged limbs on Urbilateria are unlikely. But it did have the genes for making limbs. And we know that these genes are used for making all sorts of things that project out from the body. So, even if Urbilateria didn't walk or swim, it may have had structures that projected out from the main body, perhaps things that helped it detect (e.g., a sensory apparatus) or ingest (a mouth or tentacles) food. Later in the Cambrian the genes used for making these projections would be used for making bona fide walking and swimming limbs.

If Urbilateria was certain to have had the *tinman* gene, did it have a heart? We would not expect a modern heart like ours. But it could have had some collection of contractile cells for pumping fluids around the body. In addition, the number of different *Hox* genes suggest that at least the front, middle, and back end of Urbilateria may have been distinct.

And, using gene and developmental logic, we can say that it certainly had a throughgut with a mouth and anus. We can also confidently say all sorts of cell types—muscle, nerve, contractile, photoreceptive, digestive, secretory, and phagocytic—existed because these exist in all descendants. The uncertainty about Urbilateria is the degree of organization of these cells into organs that we would call eyes, hearts, limbs, etc. The organization was complex enough to lock in the function of *Pax-6*, *Dll*, *tinman*, *Hox* genes, etc., into roles that have been preserved in all of this ancestor's descendants for more than 500 million years.

I have to be tentative here because we can't and won't know for certain until we find the fossils (and the search for new sites and types of deposits is ongoing). But the important new sketch that Evo Devo has provided is that of an animal equipped with all of the necessary genes for building complex bodies and possessing some initial level of anatomical complexity.

Darwin, in a letter to the geologist Charles Lyell, had also speculated about our ancestors. Extrapolating from comparisons of vertebrates he said, "Our ancestor was an animal which breathed water, had a swim bladder, a great swimming tail, an imperfect skull, and undoubtedly was an hermaphrodite! Here is a pleasant genealogy for mankind." The discoveries of similarities stretching across most of the animal kingdom allow us to glimpse even further into deep time for an earlier creature, a beast that makes Darwin's ancestor seem downright sophisticated.

Be proud of your heritage.

The Cambrian Explosion: So Many Arthropods, So Little Time

The geological boundary of the beginning of the Cambrian period has been dated to 543 ± 1 million years ago. But that boundary does not mark the overt record of animal evolution. Relatively few forms appear for the next 15 to 20 million years in the fossil record until definitive arthropods, chordates, echinoderms, and brachiopods arrive. Since these forms are well differentiated from one another (they have to be in order to make the classification), it is inferred that the diversification of various animal lines was under way for a considerable time, albeit cryptically with respect to the fossil record.

Simon Conway Morris, one of the leading paleontologists deciphering the events of the Cambrian, has likened this early phase of diversification to a trail of gunpowder leading back into the "mists of time." Whatever the length of this trail, by the late Early Cambrian, it reached the powder keg and the diversity of forms exploded. This is not just the appearance of individual representatives of major groups, but a parade of variations on basic body types. The Burgess Shale alone contains about 140 species of animals representing more than ten phyla. Other sites have yielded additional bounty; in particular, the Chengjiang fauna found in the Yunnan province of China is notable for spectacularly preserved specimens and, perhaps even more important, it is about 15 million years older than the Burgess Shale. The Chengjiang fauna helps push back the earliest known appearance of several groups. It is a very active site continuing to produce finds that wow us, including some fantastic vertebrates I will discuss shortly. The Chengjiang also affords a snapshot comparison of Cambrian life at another time interval, in another region of that ancient world. The Chengjiang along with the Burgess tells us that two groups in particular diversified in spectacular fashion, the arthropods and the *lobopodians*. Named after their simple unjointed legs, the lobopodians are less well-known, but they play a crucial role in the story of arthropods and of the Cambrian in general.

All of the Cambrian faunas are dominated by arthropods. Fully one-third or more of all Burgess species are arthropods. They range from the familiar trilobite forms, such as *Olenoides serratus* (on page 136) or *Naroaoia compacta*, to the less familiar types of *Waptia fieldensis*, *Marrella splendens* (the most abundant Burgess fossil), and

Canadaspis perfecta (figure 6.3). In all of these animals, a dominant feature is the similar appearance of many segments and their associated appendages. In fact, this very repetitive organization is not at all restricted to the arthropods—the lobopodians also display large numbers of relatively few different types of body parts.

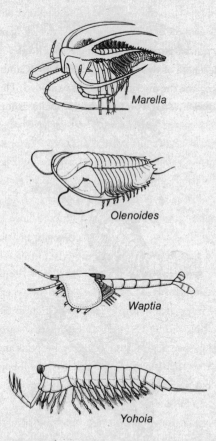

Marella

Olenoides

Waptia

Yohoia

FIG. 6.3 **Cambrian arthropods of the Burgess Shale.** These animals display different variations on the arthropod design, with different numbers and kinds of specialized jointed limbs. DRAWING BY LEANNE OLDS

Among the lobopodians were some of Gould's favorite creatures, including an animal called *Aysheaia* (figure 6.4). One reason for the acute interest in *Aysheaia* was its simple repetitive body organization and tubelike limbs. These features raised the possibility that this animal represented a primitive form, a precursor of more elaborate body

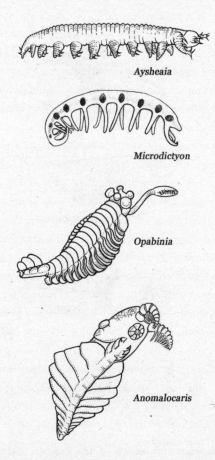

Aysheaia

Microdictyon

Opabinia

Anomalocaris

FIG. 6.4 **Cambrian lobopodians of the Burgess Shale.** These animals have unjointed limbs and are the closest relatives of the arthropods. DRAWING BY LEANNE OLDS

and limb patterns. Walcott first thought that *Aysheaia* was an annelid worm. Others, including Gould, believed rightfully that it belongs to the lobopodians. This group is most closely related to a modern phylum of soft-bodied multilegged critters called Onychophora (or velvet worms). Importantly for my story, the living Onychophora and extinct fossil lobopodians are the closest relatives to arthropods, a so-called sister group. It is believed that arthropods evolved from some lobopodian-type ancestor. Lobopodian fossils have been especially informative with respect to developing a picture of how the ancestral arthropod body and limb design evolved.

Some of the most spectacular Cambrian animals are lobopodians that sit fairly close, morphologically speaking, to what some paleontologists believe was the primitive form of arthropods (figure 6.4). Detailed study of fossils such as *Opabinia* and the frightening *Anomalocaris*, along with other lobopodians and arthropods, suggests that a series of innovations took place in lobopodians—including segmentation, a hard exoskeleton, and a biramous, or forked, limb—and became basic features of all arthropods. Various lobopodians possessed some subsets of these characters, ranging from *Aysheaia*, which had none of them and is viewed as primitive in the group, to *Opabinia* which was segmented but lacked fully biramous limbs, to *Anomalocaris*, which had biramous limbs but did not have a fully hardened exoskeleton (I'll have a lot more to say about the tremendous opportunities created by this basic limb design in the next chapter).

The variety of lobopodians and arthropods in the Chengjiang, Burgess, and other sites allows us to view the Cambrian not as an instantaneous event—a sort of "now you don't see 'em, now you do" magic act—but as a substantial episode during which body design was evolving. Ten or 15 million years is rapid in the larger scheme of Earth's and life's history, but it is *plenty* of time to invent new appendages, change body design, etc. (For comparison, realize that most kinds of mammals—primates, rodents, bats, shrews, carnivores, etc.—appear in the fossil record within the first 10–15 million years after the disappearance of the dinosaurs 65 million years ago.)

The question for us here, though, is: What drove this evolution? On this, Evo Devo has new insights to offer.

New Genes for New Animals?

The simplest and, for a long time, the most commonly held idea relating genes to the evolution of complex form is that new genes must evolve in order for new kinds of body designs and structures to arise. The intuitive appeal of such a notion is understandable. Since the form of a given species is due to its unique genetic information, then new forms require new information—ergo new genes. But as we will soon see, despite its appeal, the invention of "new genes" is not the explanation for the origin or diversity of most animal groups.

The first version of the "new genes" idea related to any specific group of animals was put forth for the arthropods by Ed Lewis at Caltech, based on his pioneering Nobel Prize–winning studies of fruit fly *Hox* genes. Lewis suggested that the many *Hox* genes that specified the many different body segment types in insects evolved from a smaller set of *Hox* genes that specified the smaller number of distinct segment types in early insect and arthropod ancestors. The Lewis hypothesis was incorrect. But the testing of this idea illustrates very nicely how the logic of Evo Devo works and has led to a clear picture of how the many kinds of arthropod bodies did evolve.

How can we learn about the genes of arthropod ancestors? The strategy hinges again on the critical logic that whatever is shared by two or more groups is likely to have existed in their common ancestor. But what animals can we study? *Opabinia*, *Anomalocaris*, and their Cambrian brethren are long gone. True, but animals bearing lobopods have not disappeared entirely. The Onychophora not only look somewhat like the

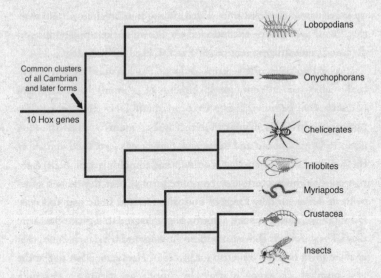

FIG. 6.5 **Evolutionary tree of arthropods and lobopodians.**
The relationships of living and extinct groups are shown. The
common ancestor of both groups that most likely lived prior to the
Cambrian must have possessed at least ten *Hox* genes because that
number is found in all living descendants. DRAWING BY JOSH KLAISS

Burgess fossil *Aysheaia*, but they still walk the Earth on lobopods just as
their Cambrian ancestors did (figure 6.5). My students Bob Warren, Jen
Grenier, and Ted Garber and I figured that Onychophora would be our
best shot at asking about the genes of arthropod ancestors, because any
genes shared between living Onychophora and living arthropods had to
exist in their last common ancestor.

The problem for us was, there are no Onychophorans living in the
United States, let alone in Wisconsin. But Australia has a lot of them, so
I "forced" Bob and Jen to leave Wisconsin in the middle of our beautiful
winter to go to New South Wales, Australia, where our collaborator and
Onychophoran-hunting expert Paul Whitington (then at the University
of New England in Armidale, NSW) would show them where to find

these reclusive and well-camouflaged creatures in fallen logs. "No worries," Paul told them—except mind the brown snakes, poisonous spiders, and giant stinging centipedes that also liked the same logs!

It took two collecting seasons to do so, but Jen and Bob were eventually able to find enough specimens of a small brown species (*Akanthokera kaputensis*; figure 6.6) to obtain DNA and embryos for further study. Their primary interest was to identify all of the *Hox* genes in Onychophora and to see how these genes were used in making these animals. We knew that fruit flies had ten *Hox* genes in all, eight conventional ones and two not so conventional ones that served other roles in development. The key question for Jen, Bob, and Ted was: How many *Hox* genes did Onychophora have and which ones were they? They isolated DNA from these animals and used techniques that enabled them to selectively fish for pieces of *Hox* gene DNA out of the vast number of genes in the Onychophoran genome. Although Onychophorans have only a few distinct types of segments and appendages, our team found that, despite their relative simplicity, they had all of the *Hox* genes known from flies and other arthropods.

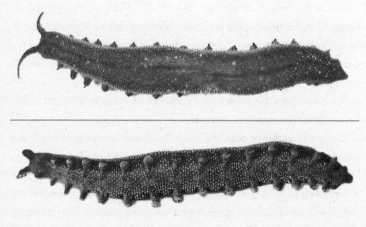

FIG. 6.6 An Onychophoran, *Akanthokera kaputensis*. PHOTO BY JEN GRENIER AND STEVE PADDOCK

This tells us that all of the arthropod *Hox* genes were in place in the last common ancestor of Onychophorans and arthropods. And that means that all of those Cambrian lobopodians and arthropods also had the same large set of ten *Hox* genes, from *Aysheaia* to *Anomalocaris*, *Microdictyon* to *Marrella*. Furthermore, the bodies of all of the later arthropod designs—spiders, centipedes, insects, and all sorts of crustaceans—were sculpted by the same set of *Hox* genes.

Right up to the first report of our findings, the notion that the Cambrian might have been ignited by an expansion in the number of *Hox* genes in animals was favored by many paleontologists. Although our data nixed this completely, this was not a disappointment whatsoever. The ability to actually test this idea by examining the genes of obscure creatures illustrated the new power of Evo Devo to shed light on the distant past. For "indoor" molecular biologists like myself and my students, it was a thrill to have something meaningful to contribute to the story of the Cambrian.

But this was just the beginning. The question remained: If not new *Hox* genes, then how did Cambrian and later forms evolve? The mere presence of certain sets of genes in DNA did not tell us the answer. The key was to look at embryo geography and the making of different kinds of arthropods. They tell us that the evolution of form is not so much about what genes you have, but about how you use them.

Shifty *Hox* Genes and Williston's Law

The story of arthropod evolution from the Cambrian onward is largely one of increasing segment and limb type diversity. Trilobites possessed three main body regions—head, trunk, and pygidium—and in each of them, most or all of the segments and appendages appeared very similar to one another, differing generally only in size. In the living groups of arthropods, representatives of which had all appeared before or within the first 150 million years after the end of the Cambrian, the range of appendage types is much greater, to as many as a dozen or more. Appendages on the head, trunk, and tail of arthropods have become specialized for feeding, locomotion, respiration, burrowing, sensation, copulation, brooding young, and defense. The great success of the arthropods is no doubt due to the adaptations brought about by the increased specialization of limb types.

How did the number of distinct appendage types increase? There must have been considerable changes in the geography of arthropod embryos. To understand what happened, we again look to living animals. The genetic control of appendage type is best known in fruit flies, where we know that the formation of each kind of appendage—a variety of jaw appendages, each of the three different pairs of legs, the (generally) limbless abdomen, and the genitalia (also a modified appendage)—is controlled by Hox proteins. This great diversity of appendage types and functions has been achieved by deploying different *Hox* genes in different zones along the main body axis. The geography of the insect embryo involves the generation of many unique individual or combined *Hox* "zones" (the genes' zones are depicted with the numbers 1 through 10 in figure 6.7).

What was the geography of Cambrian embryos? How were *Hox* genes deployed in these animals 500 million years ago? We can't look directly, but we again can make inferences based on comparisons of embryo geography and how *Hox* genes are used in different living arthropods. For example, some arthropods such as brine shrimp have a very simple thorax in which all segments and appendages are very similar. This is also thought to have been the general geography in its primitive ancestors. Figure 6.7 shows how in brine shrimp embryos, two Hox proteins (numbers 8 and 9) are expressed in virtually identical patterns in the developing thorax, not in different zones as in insects. In centipedes, a different major group of arthropods, the geography of *Hox* genes in the embryo is similar to that in the brine shrimp. The long trunk of the body is made up of identical segments bearing identical limbs. In centipede embryos, the same two Hox proteins (numbers

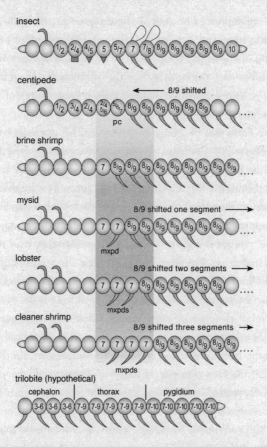

FIG. 6.7 **Shifting zones of *Hox* gene expression shape the major differences in arthropod design.** *Hox* genes are indicated by number. Note the relative shifts in the position of boundaries of *Hox* genes 7, 8, and 9 among insects, centipedes, and four types of crustaceans (brine shrimps, mysids, lobsters, and cleaner shrimps) (shaded area). The number of maxillopeds (mxpds) correlates perfectly with the number of segments that express genes 8/9, and is shifted rearward from the condition in brine shrimps (which have no maxillopeds). The centipede has a poison claw (pc) just before its legs. Trilobites probably had just three body zones, defined by three different combinations of *Hox* genes. DRAWING BY LEANNE OLDS

8 and 9) are expressed in each of these segments and limbs. In both arthropods, a zone of identically patterned segments reflects a zone of expression of the same Hox protein or proteins. Therefore, it follows that in Cambrian arthropods such as the trilobites, it is most likely that blocks of similar segments and appendages reflected zones of the same Hox proteins.

We also know that the boundaries between *Hox* expression zones are usually reflected by a change in segment and appendage type in arthropods. In the brine shrimp and centipede, the segment just forward of the long thorax expresses a different Hox protein or combination of Hox proteins (numbers 7 and 5/6/7, respectively) and forms a different type of appendage. In the shrimp, this is a feeding appendage; in the centipede, this appendage will become the poison claw used to disable prey and in defense. The relationship between differences in appendage type and *Hox* zones up and down the body axis is widespread.

The picture across the arthropods is that shifting zones of Hox proteins correlate closely with evolutionary differences in the number and kind of appendages projecting from segments. This is not only the case between the major classes of arthropods, but even within them. The role of shifting *Hox* zones in evolution was beautifully demonstrated by Michalis Averof and Nipam Patel, who collected and examined embryos from a wide variety of crustaceans (the group of arthropods that includes shrimps, barnacles, crabs, and lobsters). One of the prominent differences between groups was in the number of maxillipeds, the feeding appendages at the front end of the thorax that have been modified from limbs. Brine shrimp do not have maxillipeds, nor did primitive crustaceans. Other crustaceans, however, have one, two, or (as in lobsters) as many as three pairs of maxillipeds. Small changes in embryo geography underlie these important differences in crustaceans. Averof and Patel found that the zones of expression of two Hox proteins (numbers 8 and 9) were shifted backward by 1, 2, or 3 segments in these crustaceans, respectively, relative to those crustaceans without maxillipeds (figure 6.7). The extent of the shift correlates perfectly with the number of maxillipeds. Furthermore, these shifts and maxillipeds appeared to have evolved several times independently in crustacea, suggesting that similar functional adaptations were achieved by similar mechanisms in different animals. (I'll say more about the significance of repeated instances of similar changes in the next chapter.)

Shifting *Hox* zones have sculpted the prominent differences along the main body axes of living arthropod groups such as spiders, crustaceans, centipedes, and insects. It is a very reasonable extrapolation to assert that this was also the story in the Cambrian, where body regionalization and appendage specialization are evident in all fossil arthropods. The blocks of similar segments in fossil taxa were certainly the zones of particular *Hox* genes (figure 6.7). The increase in the number of different appendage and segment types in arthropod evolution is the product of generating a greater number of unique zones in the embryo in which specific individual or combinations of *Hox* genes are expressed. *This relative shifting of* Hox *zones is therefore one of the mechanisms underpinning Williston's Law*—the specialization of repetitive parts requires that the different parts fall into different *Hox* zones.

Shifting *Hox* zones is not just an arthropod phenomenon—this same primary mechanism underlies major features of the anatomical diversity of our own phylum, the vertebrates.

The Making of the Vertebrates: More *Hox* Genes and Many Shifts

Our family line is also traceable back as far as the Cambrian. We are vertebrates, part of a larger group of animals known as chordates that possess a notochord. The chordates also include tunicates (such as sea squirts) and cephalochordates such as the lancelet. Chordates are part of the deuterostome branch of the animal tree (figure 6.8). The Burgess fossil *Pikaia* was for a long while the best-known ancient chordate but

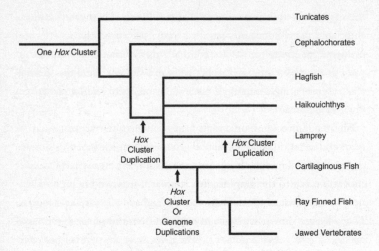

Fɪɢ. 6.8 **The chordate evolutionary tree and the expansion of *Hox* clusters in vertebrate evolution.** The common ancestor of all chordates had one cluster, as do living tunicates and cephalochordates. Cluster duplication has happened several times since, on the line to jawless fish, on the line to cartilaginous fish (sharks), and again in lampreys. Because *Haikouichthys* is a Cambrian vertebrate whose evolutionary relationships are not certain, on the tree it branches out at the same (unresolved) time as hagfish, lamprey, and cartilaginous fish. DRAWING BY JOSH KLAISS

spectacular recent finds in the Chengjiang have pushed back the earliest appearance of vertebrates to about 520 million years ago and a treasure trove of some species has revealed details of a surprisingly complex anatomy for vertebrates at this time.

Specimens of the fossil jawless fish *Haikouichthys ercaicunensis* reveal the presence of a head lobe with eyes, possibly nasal sacs, ten or more separated vertebral elements, gills, a dorsal fin, and a ventral fin. This anatomy is more complex than the later *Pikaia* and indicates that the evolution of the vertebrate body was well advanced by the Early Cambrian. These recent discoveries underscore the tremendous importance of the fossil record and of ongoing and new excavations. The date of the first appearance of groups and physical features is always tentative because subsequent finds can always push it back—in this case a critical 15 million years earlier. Furthermore, while vertebrates were not the most numerous group in the Early and Middle Cambrian, their discovery in Chengjiang puts these predatory *Haikouichthys* firmly in the picture of Cambrian ecosystems.

The invention and modification of many structures marked the origin of the vertebrates, including much more complex brains, sensory structures, cartilage, the body skeleton, and skull. Many subsequent innovations led to the amphibians, reptiles, birds, and mammals we know so well. Just as for the lobopodians and arthropods, we'd like to know whether the early evolution of the vertebrates in the Cambrian depended upon a very similar tool kit of developmental genes as that then possessed by other groups, or whether some changes in the tool kit might have played a role in the origin of ancestral vertebrates.

We cannot, of course, recover genes from *Haikouichthys*, but we can study some proxies, living species that occupy key places on the chordate and deuterostome family trees, to allow us to infer the genetic complexity of ancestral vertebrates. One key group is the cephalochordates. These animals lack the vertebrate features of a cranium or bony structures, but they are the sister group to vertebrates in the same way that living Onychophorans are the sister group to arthropods. The content of the *Hox* cluster in cephalochordates reveals that cluster's content in the last common ancestor of cephalochordates and vertebrates.

The lancelet is the one and only cephalochordate still around today. These two- to three-inch-long animals can be found in Tampa Bay, Florida, and some other waters. When Jordi Garcia-Fernandez and Peter Holland examined its *Hox* genes, they found just a single cluster of *Hox* genes. Recall that modern vertebrates such as mice and ourselves have four *Hox* clusters containing thirty-nine genes in all. The lancelet tells us that an increase in *Hox* cluster number happened sometime after the split between the vertebrate and cephalochordate lines, in the Cambrian or a bit earlier. We also know that other deuterostomes such as tunicates and echinoderms possess single *Hox* clusters. So, while the entire diversity of tunicates and echinoderms in the Cambrian and ever since has, like the arthropods, evolved around a cluster containing ten or so *Hox* genes, the vertebrates did expand their number of *Hox* genes.

When in vertebrate evolution did the number of *Hox* clusters increase? Could this increase have triggered vertebrate evolution? To answer these questions, a whole bunch of living species representative of different groups on different branches of the vertebrate family tree had to be examined. All mammals, birds, and certain fish, including the primitive, deep-sea-dwelling coelacanth, have four *Hox* clusters. It is safe to conclude then that there were four *Hox* clusters in the common ancestor of all of these jawed vertebrates.

But more primitive living vertebrates such as lampreys have fewer *Hox* clusters. Detailed scrutiny of these clusters' genes and comparison with those of bony fish and mammals suggests that our four *Hox* clusters are the product of two rounds of *Hox* cluster duplication early in vertebrate evolution; one round after the split between cephalochordates and the origin of lampreys and again sometime before the origin of bony fish. Looking at our family tree again (figure 6.8), we see that because the Chengjiang fossils are jawless fish, we infer that they could have had just one or perhaps two *Hox* clusters.

It turns out that the different numbers of *Hox* clusters in different vertebrates reflects differences in the overall size of the whole genetic tool kit. Not only were *Hox* gene clusters duplicated in vertebrate evolution, but many other kinds of genes in the tool kit were as well. One way this might have happened is through duplications of the whole genome, or large sections of it. The expanded tool kit in higher vertebrates tells us that for the early history of vertebrates, there is certainly strong support for the idea that the evolution of more genes played a role in the evolution of body design. One index of the anatomical evolution of chordates is the number of cell types in different groups. Humans and other higher vertebrates have many more cell types than cephalochordates, which lack the kinds of cells that give rise to our cartilage, bones, head, and certain sensory structures. This means that one correlate of a greater number of tool kit genes is greater cell type and tissue complexity—achieved by using more genes to generate more combinations of developmental instructions.

However, more genes is not the main story for the subsequent evolution of higher vertebrates. It is crucial to remember that the presence of four *Hox* clusters has been stable throughout the evolution of amphibians, reptiles, birds, and mammals. Frogs and snakes, dinosaurs and ostriches, giraffes and whales, have evolved around a similar set of four *Hox* gene clusters. So again, the mere number of *Hox* genes does not tell us how these forms evolved. The making of the diversity of these animals with respect to their main body axes and body parts is, as in the arthropods, one of changing embryo geography by shifting the zones of *Hox* genes, albeit a larger number of them.

For example, in vertebrates the transition from one vertebral type to another—cervical/thoracic, thoracic/lumbar, lumbar/sacral, sacral-caudal—corresponds to transitions between the zones of expression of particular *Hox* genes. The forward boundary of expression of one of these genes, *Hoxc6*, falls at the cervical/thoracic boundary in mice, chickens, and geese, even though each of these animals has a different number of thoracic vertebrae. Thus, relative to vertebral number, the relative position of *Hoxc6* expression has shifted among these animals (figure 6.9). In snakes, the shift is even more dramatic. There is no clear cervical/thoracic boundary in snakes and *Hoxc6* expression extends way up into the head. All of these vertebrae bear ribs, which indicates these are of a thoracic type, but they also have some features of neck vertebrae, suggesting that snakes got longer by losing their neck and extending their thorax via shifts in *Hox* zones.

It is striking, and very satisfying, to discover that the evolution of body forms in two of the most successful and diverse groups of animals—arthropods and vertebrates—has been shaped by similar mechanisms of shifting *Hox* genes up and down the main body axis.

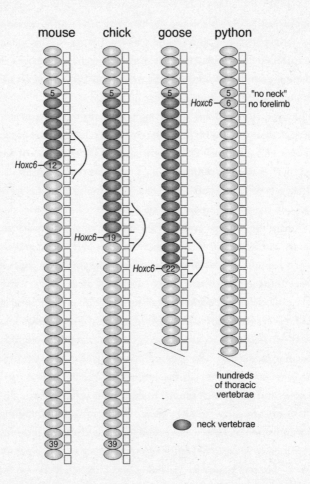

mouse   chick   goose   python

Hoxc6—12

Hoxc6—19

Hoxc6—22

Hoxc6—6   "no neck"
no forelimb

hundreds
of thoracic
vertebrae

neck vertebrae

FIG. 6.9 **Shifting zones of *Hox* expression also shape vertebrate diversity.** Different vertebrates have different numbers of neck vertebrae, with mice having a short neck, geese long necks, and pythons virtually no neck at all (just a long trunk). The boundary between neck and trunk vertebrae is marked by expression of the *Hoxc6* gene in all cases, but the position differs in each animal relative to the overall body. The forelimb arises at this boundary in all four-legged vertebrates; in snakes this boundary is shifted far forward to the base of the skull and no limbs develop. DRAWING BY LEANNE OLDS

The great message here is that an understanding of many of the large-scale changes in animal design is within our grasp. We can begin to think of individual groups—insects, spiders, and centipedes, or birds, mammals, and reptiles, as well as their long extinct fossil relatives—not so much in terms of their uniqueness, but as variations on a common theme. Denis Diderot, the brilliant writer-philosopher of the second half of the eighteenth century, captured the essence of this in his observation quoted at the beginning of this chapter, just as Williston did in formulating his law nearly a century ago. Now we have a general mechanism for, and thus a very tidy explanation of, one of the great trends in animal evolution.

Switches Make the Shift

Let's take this understanding one more step to a deeper level, beyond *Hox* genes and embryo geography, to ask how these shifts in *Hox* zones and changes in anatomy occur.

Switches. It is the genetic switches of *Hox* genes that control the coordinates of *Hox* zones in embryos. Evolutionary shifts in *Hox* zones arise through changes in the DNA sequences of *Hox* gene switches.

For example, the vertebral column of the mouse consists of seven cervical vertebrae and thirteen thoracic vertebrae, whereas the chicken vertebral column consists of fourteen cervical and seven thoracic vertebrae. The forward boundary of the *Hoxc8* gene expression zone is farther back in the chicken embryo than in the mouse embryo. The boundaries of *Hoxc8* expression in the early mouse and chicken embryos are controlled by a specific switch. Changes in the DNA sequence of this switch between the mouse and chick are responsible for the differences in the relative position of *Hoxc8* expression in the two species.

The evolution of the *Hoxc8* switch between these two vertebrate classes illustrates a crucial general point about the role of switches in evolution. Changing the sequence of switches allows for changes in embryo geography *without disrupting the functional integrity of a tool kit protein*. In this case, changing the *Hoxc8* switch enables changes in the number of particular types of vertebrae. The Hoxc8 protein plays crucial roles in other tissues, so mutations in the coding sequence of the gene are likely to affect all of its functions. Changing a specific switch enables specific modules to change without affecting other body parts.

This is the same strategy underlying the changes in crustacean or other arthropod geographies I described. Shifting *Hox* expression backward by 1, 2, or 3 segments is a matter of changing switches to activate *Hox* genes at slightly different coordinates, without disrupting the function of the Hox protein.

Rethinking the Cambrian: Genetic Possibilities Find Ecological Opportunity

The new perspective from Evo Devo on the early history of animals has three important elements. First, it suggests that the last common ancestor of the two main branches of the animal tree was fairly complex, genetically and anatomically, in spite of its paleontologically cryptic existence in the Precambrian. Second, we know for certain that the full genetic tool kit for body-building was in place, but its potential was largely untapped for a considerable length of time. And third, the potential of the tool kit was realized largely through the evolution of switches and gene networks and the shifting of *Hox* zones, in the Cambrian and more recent periods.

If the invention of tool kit genes per se was not the trigger of the Cambrian Explosion, then what was? It is becoming increasingly appreciated that the Cambrian Explosion was an ecological phenomenon. The evolution of larger, more complex animals, once under way, led to the evolution of still larger, more complex animals, and so on. As the Big Bang unfolded, the pressure of ecological interactions and competition among increasingly diverse animal species drove the evolution of more complex structures—compound and camera eyes for vision; jointed appendages for walking, swimming, and grabbing prey; hearts for managing circulation in larger bodies; and refinement of the body into head, trunk, and tail sections that facilitated more complex movements and defense. Genes in the tool kit are important actors in this picture, but the tool kit itself represents only possibilities, not destiny. The drama of the Cambrian was driven by ecology on a global scale.

In the Cambrian, the diversity of many different animal groups expanded from a small initial state. Many other expansions, or "Little Bangs," have occurred since the Cambrian and can often be attributed to the exploitation of new ecological opportunities. The invasions of the land by vertebrates and arthropods (and plants) were followed by explosive expansions. In many cases, these invasions were enabled by structural innovations, changes in the geography of individual body parts that opened up a whole new way of living, and set the stage for yet further expansions. In the next chapter, we'll look at a few key innovations that "made" whole new types of animals.

The evolution of animal appendages has played a central role in the origin and expansion of specific groups depicted here.

7

Little Bangs: Wings and Other Revolutionary Inventions

I'm learning to fly, but I ain't got wings
Coming down is the hardest thing

—Tom Petty and Jeff Lynne,
"Learning to Fly" (1991)

ON THE RARE OCCASIONS when I sit down for a meal at a very fancy restaurant, I am, like many people, intimidated by the cutlery. Which again is the salad fork, and my main dinner fork? Uh-oh, did I just eat French fries with my dessert fork? Butter knives, steak knives, cheese knives, tablespoons, teaspoons, soup spoons—how did we get to this state of overspecialization?

Dining etiquette was definitely simpler in the Middle Ages, but evolving. People began to eat with two knives, using one to cut the food and a second to spear it and bring it to their mouths. The fork then appeared as a two-tined prong that was much more efficient than a single-bladed knife in picking up food.

It isn't precisely known when and where forks started replacing the second knife at the dining table, and goodness knows what perverse minds came up with all of those other utensils, but this little snippet of cutlery history is analogous to a broad trend in biological evolution. Namely, structures (like the fork) that evolve a dedicated function (spearing food) are often derived from a preexisting structure (the knife) that served more than one role (cutting and spearing). The duplication of the original structure (the practice of using two knives) enabled the subdivision of labor among two distinct structures. Furthermore, selected for a new purpose, the structure can then evolve further modifications and specializations.

The history of the humble paper clip provides another lesson from everyday life about the great affairs of evolution. It was first invented to compete with the pin in holding fabric together; only later did its principal use become to hold papers. Many variants appeared between the first clip and the design in widest use today. Some of these variants were specifically for holding newspapers, others were designed for holding larger reams of paper (figure 7.1). The evolution of the paper clip illustrates how one structure, invented for a particular use, can evolve new forms and adapt to new functions.

The histories of cutlery and paper clips are analogous to the evolution of animal appendages. Freed of some tasks, subsets of appendages have evolved new forms and functions that enabled species to compete in an intensely competitive natural world. First in the oceans, then on land, and subsequently in the air, the continuing drama of evolution has been an "arms race"—or, more literally, a "limbs race"—to find better, faster, lighter, stronger, or more nimble limbs with which to

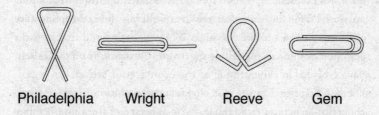

FIG. 7.1 **Evolution of the paper clip.** An example of a form adapting for better performance. DRAWING BY LEANNE OLDS

swim, walk, run, hop, breathe, burrow, or fly, or to grab, crush, swallow, poke, filter, suck, or chew food.

Often, these inventions opened up whole new ways of living that led to rapid expansions of diversity—"Little Bangs" of animal evolution. Some innovations were followed by further evolutionary changes as opportunities were exploited. Vertebrates came to land on the modified pectoral and pelvic fins of fish ancestors. Although they had just two pairs of limbs to work with, they have taken to the air three separate times to give rise to new types of animals (pterosaurs, birds, and bats), returned to the water many times (whales and dolphins, seals, etc.), and evolved all sorts of limbs for making way on land. Several million years ago, when human ancestors stopped walking on their knuckles and stood upright, new opportunities arose for our forelimbs. Freed of the burden of carrying the weight of the body, our arms and hands could be used for all sorts of activities—toolmaking, hunting, communication, and eventually the production of symbolic records of the natural world. These activities were supported by, and in turn drove, the evolution of bigger and faster brains, which required further evolution of skeletal anatomy for bearing offspring, and family structures for extended periods of parental care.

The importance of serially repetitive body design is the ability to shift the burden of some task from two or more pairs of structures onto fewer structures, then to specialize the freed-up structures for new purposes. This has served vertebrates well, but the arthropods have absolutely run riot with the idea. While *all* arthropod limbs have a common design, an incredible spectrum of variations on this design have evolved. In the example of the evolution of crustacean maxillipeds I described in the last chapter, the evolution of these food-collecting structures freed the thoracic appendages from filter-feeding duty and allowed their adaptation to new modes of locomotion, such as walking, swimming, and burrowing. This in turn opened opportunities that were exploited in a burst of crustacean evolution.

The key insight from the diversification of serially repeated structures, especially limbs, is how evolutionary transitions are accomplished. One of the long-standing challenges in evolutionary biology has been to understand how major changes occurred in the distant past. The erroneous notion of the anti-evolution camp has been that the intermediate stages in the evolution of structures must be useless—the old saw of "What use is half a leg or half an eye?" Following this preposterous "logic," the conclusion is that structures must be forged perfectly in one instant—that evolution didn't happen. This view clutches desperately at Darwin's own explicit discussion of difficulties with the theory of natural selection in the *Origin*, yet it always fails to grasp or cite Darwin's brilliant resolution of the matter. The crucial insight he had was that the same organ often performs wholly distinct functions at the same time, and that two distinct organs may also simultaneously contribute to the same function. This multifunctionality and redundancy create the opportunity for the evolution of specialization through the division of labor. The availability of duplicate structures enables animals to "have their cake and eat it too" or, more accurately, to "have their limbs while learning to eat with them too."

In this chapter, I will highlight the power of Evo Devo to reveal the continuity among structures that have been adapted to vastly different purposes, particularly in the arthropods. Because this continuity was often concealed by differences in form, biologists were previously uncertain about the relationships between different structures in different groups, such as the gills of aquatic crustaceans, and the appendages of terrestrial arthropods. But thanks to new methods and insights from Evo Devo, this uncertainty has vanished. I will focus on how the simple tube-like walking limbs of Cambrian lobopodians became efficient articulated swimming, walking, and respiratory appendages in crustaceans, gills in aquatic insects, wings in terrestrial insects, and book lungs and spinnerets in spiders. None of these later structures were invented from scratch; they are all variations on an ancient limb design.

The remodeling of limbs to new forms and functions is a matter of changing the geography of limb development. I will show how changes in geography enabled insects to fly and new modes of flight to evolve, and enabled vertebrates to come onto land, and snakes and other groups to adapt to new niches.

From So Simple a Beginning: The Biramous Limb

In the arthropods, from a very simple beginning, a dazzling variety of versatile appendages have evolved from an ancestral limb design. Many different appendages co-occur in individual species that provide both tool-like dexterity as well as savage weaponry. Take a look at all of the different implements a humble crayfish carries (figure 7.2)—there are more gizmos on this one animal than on a deluxe Swiss Army knife.

Limb morphology has figured prominently in many debates about arthropod evolution. New light on the origin and evolution of structures has been thrown by a combination of sources. Paleontology has uncovered and interpreted key fossils, advances in the study of animal relationships have sorted out some of the branches of the arthropod tree, and Evo Devo has provided an entirely new and decisive kind of evidence.

The whole story of arthropod limb evolution revolves around the origin and modification of an ancestral biramous (forked) limb. The

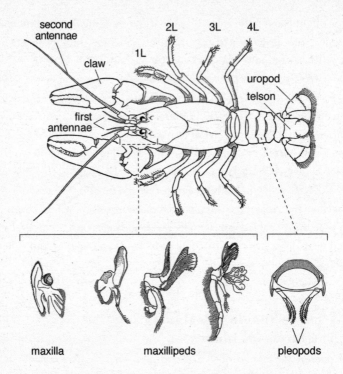

F IG . 7.2 **The great diversity of appendages in a single arthro-pod, the crayfish.** Fourteen or more different kinds of appendages project from this animal, including two sets of antennae, four sets of legs, three sets of maxillipeds, and several pairs of pleopods (abdominal swimming appendages). All of these appendages are derivatives of a common ancestral design. DRAWING BY LEANNE OLDS, BASED ON DRAWINGS IN R. E. SNODGRASS'S *ARTHROPOD ANATOMY* (COMSTOCK PUBLISHING ASSOCIATES, 1952)

basic elements of this limb design, which are seen in trilobites and crustaceans, are shown schematically in cross section in figure 7.3. The two main branches of the limb fork from a common base. The inner branch forms the jointed walking leg, and the outer branch has many different roles. All sorts of smaller branches and extensions with spe-

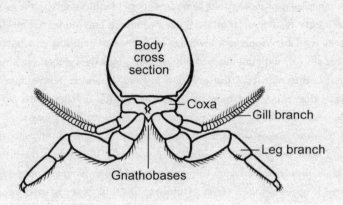

FIG. 7.3 **The biramous limb design.** End-on view of a typical branched limb where the upper gill branch is involved in respiration and the lower branch engages in locomotion. DRAWING BY LEANNE OLDS, BASED ON DRAWINGS IN S. J. GOULD'S *WONDERFUL LIFE* (W. W. NORTON, 1989)

cialized functions have evolved that project from the base, the inner, or the outer branches in various species. For example, projecting from the base are structures specialized for food handling, and on the outer branch there are gills through which aquatic arthropods exchange oxygen and carbon dioxide.

Chances are the arthropod limbs with which you are most familiar are the jointed, unbranched legs of insects. Remarkable as they are, they are among the simplest structures in the vast repertoire of arthropod appendages. So simple, in fact, that many biologists were long misled in believing that the simple legs of insects, centipedes, millipedes, and Onychophorans linked all these animals together in one group that was separate from crustaceans, trilobites, scorpions, and horseshoe crabs, all of which had fancier, forked appendages. This indeed was still one traditional view at the time of Gould's writing of *Wonderful Life*.

But, it is wrong. I tell you this not because I want to stuff your brain with incorrect ideas, but to give some sense of how misleading the sole reliance on morphology has sometimes been in understanding the evolutionary history of animals. Some prominent and influential biologists with encyclopedic knowledge of arthropod anatomy concluded that animals with unbranched (uniramous) limbs were something altogether different from those with branched (biramous) appendages, such that they must have invented their limbs independently and belonged in different phyla, not just the single group Arthropoda.

Contrary to earlier ideas about the separate origins of unbranched and branched limbs, the weight of evidence indicates that the biramous limb is evolved from the simpler tubelike lobopods of lobopodians. It appears that a series of innovations took place in arthropod ancestors starting with the simple lobopods of animals such as *Aysheaia*. Animals with lobopods projecting from the lower parts of their segments appear to have evolved separate upper lobes that may have served gill-like functions, as seen in species such as *Opabinia*. Later, the two upper and lower appendages became fused at their bases to form a segmented and jointed walking leg in arthropod ancestors (figure 7.4). One of the newer sources of support for this picture also comes from Evo Devo. The lobopods of Onychophorans and all branches of arthropod limbs express the *Distal-less* limb-building

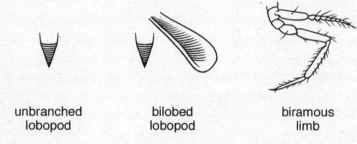

unbranched   bilobed   biramous
lobopod   lobopod   limb

FIG. 7.4 **Three stages of limb evolution.** Some lobopodians had unbranched lobopods; others had bilobed lobopods, which appear to have become fused to evolve the jointed biramous limb of arthropods. DRAWING BY LEANNE OLDS

gene. This is consistent with all arthropod limb types, both branched and unbranched, evolving from ancient lobopods, rather than the independent invention of the two kinds of limbs.

Learning to Fly:
From Many Gills to a Pair of Wings

The multifunctionality of branched aquatic arthropod limbs is key to understanding two of the major transitions that arthropods made—to walking on land and flying in the air. In aquatic crustaceans, a lobe on the outer branch of their limbs serves a respiratory function, while the inner branch is involved in swimming or walking. In all terrestrial arthropods the walking limbs are unbranched. This is a simplified condition that evolved by reducing the complex biramous ancestral appendage to just the inner branch.

This helps us understand where insect legs came from, but what about the wings? The origin of insect wings has long been a contentious mystery. Some biologists argued that wings arose as independent outgrowths of the thoracic body wall in wingless insects. A second theory is that wings were derived from a branch of an ancestral leg—in particular, from the gills of an aquatic ancestor. Comparative anatomists long wrestled with these alternatives without reaching a consensus.

But here again is where Evo Devo has stepped in with some powerful new evidence. The study of wing development in insects, principally flies, has identified a few proteins that are required for building a wing. Two such tool kit proteins are called Apterous (mutants for this gene lack wings) and Nubbin (mutants for this gene have just little nubs of wings). In order to test the theory that wings might be derived from the gill branches of crustaceans, Michalis Averof and Stephen Cohen traced how the Apterous and Nubbin proteins are expressed in the appendage of other arthropods, especially crustaceans. They found, quite strikingly, that *apterous* and *nubbin* were selectively expressed in the respiratory lobe of the outer branch of crustacean limbs. The best explanation for this observation is that the respiratory lobe and insect wing are homologous—that is, the same body part in different forms in the two animals. The only alternative possibility would be the extraordinary coincidence that, of the hundreds of tool kit proteins that could be used to make gills and wings, these two proteins were independently selected by crustaceans and insects for building these structures. The most probable scenario is that Apterous and Nubbin were used in making respiratory lobes in an aquatic crustacean ancestor of insects and have stayed on the job ever since, as this branch evolved into wings (and, as we will see later, other structures in different animals). In insects, then, the outer and inner branches of an ancestral limb have become separated, with part of the outer branch moving to the upper part of the body and evolving into a wing, and the inner branch evolving into an unbranched walking leg.

The gill-to-wing theory always had evidence in its favor (just not enough weight to settle the matter). But, if indeed insect wings came from crustacean gill branches, does this mean that some kind of crayfish or shrimp just crawled onto land and started flying? No, not at all. There were many evolutionary steps in the transition between animals that carried a set of respiratory appendages and the origin of powered insect flight on two pairs of wings as we know it today. Some of the most helpful clues to reconstructing this transition come from fossils of long extinct insects, as well as more insights from Evo Devo. My lab has had a part in this story and this is another one of those occasions where new knowledge of fossils, genes, and embryos came together to build a compelling picture.

Some of the most informative insect fossils don't look like anything around today. I was particularly struck by this when I first encountered the sketch of the "primitive aquatic nymph" fossil shown at the left in figure 7.5. Paleontologists Robin Wootton and Jarmilla Kukalova-Peck had each studied this creature, which lived more than 300 million years ago. Its most important feature is the presence of winglike structures on all of the thoracic and abdominal segments. The winglike nature of these structures is reflected in the pattern of canals in the appendages, which is similar to the general pattern of veins in winged insects. However, this fossil is of an aquatic animal; these are not wings, but gills similar to those that can be found today on the aquatic nymph stages of dragonflies and mayflies.

The most likely scenario for wing origins is that the wings of adult terrestrial insects evolved in animals that also had gills in their larval stages. Wings could then evolve as adult structures modified from gills, without dispensing with gills at all. Mayflies and dragonflies develop

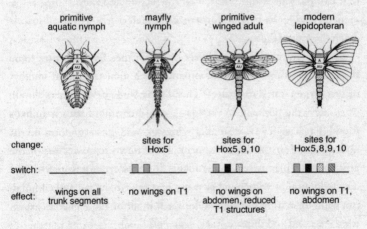

Evolution in a Switch of a Wing-promoting Gene and the Evolution of Insect Wing Number

primitive aquatic nymph	mayfly nymph	primitive winged adult	modern lepidopteran
change:	sites for Hox5	sites for Hox5,9,10	sites for Hox5,8,9,10
switch:			
effect: wings on all trunk segments	no wings on T1	no wings on abdomen, reduced T1 structures	no wings on T1, abdomen

FIG. 7.5 **Evolution of wing number and form.** Wing number has followed Williston's Law, evolving from a series of gill-like appendages on all segments of extinct aquatic nymph forms to smaller numbers and reduced structures in mayflies, to the two pairs of wings in most modern flying insects. Wing number has been progressively reduced by the evolution of progressively more sites for Hox proteins in switches of genes promoting wing development. DRAWING BY LEANNE OLDS

from immature aquatic nymphs that have gills on their abdomens, and these are the most primitive winged insects. The existence of discrete stages in animal life cycles creates a great deal of opportunity for evolution. Think about it: adult mayflies and dragonflies are entirely different animals than their aquatic young, living in entirely different environments. The adaptation to these different environments has taken place simultaneously in one genome, by separating developmental programs for building the nymph from those for building the adult. The evolution of radically different larval and adult morphologies is a pervasive theme in animals (think of caterpillars and butterflies, or bilateral echinoderm larva and pentaradial adults).

This scenario helps explain the transition of gill form to wings, but what about the number of wings? Aerodynamic studies have concluded that two pairs of wings, located on the second and third thoracic segments, provide the best performance characteristics. How did insects reach this optimal design?

Back to Williston's Law, *Hox* genes, and switches. Remember the trend that specialization is often accompanied by a reduction in the number of serially reiterated structures? This is exactly the case in the evolution of insect wing number. In the fossil record, extinct forms have been found that have fewer or smaller wings or wing-like structures on the abdomen and first thoracic segment. These forms represent intermediates between primitive aquatic forms and modern types (figure 7.5). The reduction of wing number was a matter of reducing and eliminating wing formation during development in all of the segments except for the second and third thoracic segments.

How were wings suppressed in a segment-specific pattern? The segments from which wings disappeared are zones of particular Hox proteins in all insects. Furthermore, Scott Weatherbee and Jim Langeland in my laboratory discovered that wing formation is repressed in flies in these segments by the respective Hox proteins that are expressed in them. This tells us that the modern condition of insects bearing two pairs of wings is the product of the evolution of the repression of wing formation by Hox proteins acting in the first thoracic segment and all of the abdominal segments. This repression must have evolved in stages in different groups of ancient insects, because the fossil record contains species in which wing repression was only partial. The ultimate cause of wing repression is that switches in genes involved in wing formation evolved signature sequences for Hox proteins that resulted in their being turned off in selected segments.

Spider Tales: More Adaptations of Arthropod Gills

Insects, as successful as they are, were not the only arthropods to colonize and thrive on land. Spiders have also adapted well to terrestrial life, having evolved from a different branch of the arthropods called chelicerates. They are most closely related to scorpions, mites, and horseshoe crabs.

Like insects, spiders made the transition to terrestrial life by changing the way that they breathe, move, reproduce, and capture food in comparison to their aquatic ancestors. Spiders have evolved so-called book lungs and tracheal tubes for respiration on land as well as spinnerets for producing the silk involved in web-making and capturing prey. Each of these structures form at similar positions in different segments, suggesting that they are serial homologs of one another (figure 7.6). Indeed, they each express the *Distal-less* limb-building gene during their formation, indicating that they are all modified appendages. But which appendages?

Again, Evo Devo has weighed in on a century-old question. Wim Damen, Theodora Saridaki, and Michalis Averof have found that each of these structures also expresses the Apterous and Nubbin proteins, the two tool kit proteins expressed in the gill branch of aquatic crustaceans and in insect wings. This is compelling evidence that book lungs, tracheal tubes, and spinnerets are also derived from the gill

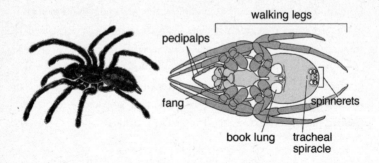

FIG. 7.6 **Spiders and their innovations.** The adaptation of spiders to land was accompanied by the evolution of book lungs, tubular tracheae, and spinnerets from the gill branches of appendages on its aquatic ancestors. All of these structures are serial homologs of the pedipalps (mouthparts) and walking legs. DRAWING BY LEANNE OLDS

branches of ancestral arthropods. Furthermore, they found that the book gills of aquatic horseshoe crabs also express this pair of proteins; these structures must also be derived from ancestral gill branches. Based upon the relationship between horseshoe crabs and spiders, these observations suggest that book gills were modified during the evolution of spiders into book lungs, tracheal tubes, and spinnerets during their adaptation to land.

Coupled with the evidence of the derivation of the insect wing from the gill branch of crustaceans, we have now a stunning picture. All sorts of terrestrial innovations are modifications of a part of an ancestral biramous limb design (figure 7.7). That design itself was most likely derived from unjointed lobopods and lateral gill lobes in lobopodians. The tools of arthropods—claws, legs, swimmerets, maxillae, maxillipeds, gills, book lungs, tracheal tubes, spinnerets, and wings—are all the products of modifying an ancestral design.

By now, you may be beginning to see a theme in evolution. Nature does not invent very often completely from scratch; rather, it remodels its existing structures with tool kit genes that are already available.

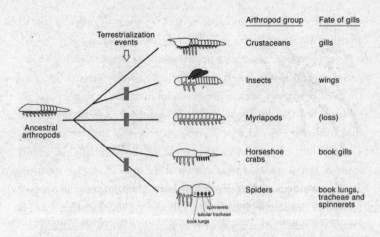

FIG. 7.7 **The many adaptations of arthropod gill branches.**
From the gill branches of aquatic ancestors, insect wings, horseshoe
crab book gills, and several spider structures evolved. This amazing
variety of structures illustrates the advantages of possessing serially
reiterated structures that can be specialized for particular tasks.
COURTESY OF MICHALIS AVEROF; ADAPTED FROM DAMEN ET AL., *CURRENT*
BIOLOGY 12 (2002): 1711; REPRINTED BY PERMISSION OF ELSEVIER

From the multiple, multifunctional feeding/swimming/respiratory/
walking appendages of aquatic arthropods, it has shaped specialized
structures that enabled species to invade entirely new ecosystems and
to establish entirely new body designs.

Evolving Appendage Geography

While all of the arthropod appendages I have described are derived
from a common ancestral design, they are not, however, identical in
form. The serially homologous spider spinnerets, tracheal tubes, and
book lungs, for example, do occur in the same animal, but these
appendages have different geographies. Furthermore, while wings also

derived from a common ancestral design, the wings of different species can differ dramatically. So, we need to be able to explain how the different geographies of serially homologous structures develop in one animal, as well as how homologous structures evolve between species.

Again, *Hox* genes are key. We know, for instance, in spiders that the segments bearing the book lungs, tracheal tubes, and spinnerets fall within different zones of *Hox* expression. This is yet another manifestation of Williston's Law. In aquatic ancestors, the book gills are serially repeated, but in spiders specialization of the appendages has occurred in adjacent segments. The developing book lungs form in the zones of *Hox* gene 7, the developing tracheal tubes in the zones of *Hox* 7 and 8, and the spinnerets in the zones of *Hox* genes 7, 8, and 9. The differences between these structures are due to the action of the different combinations of Hox proteins on switches of genes that pattern appendages.

The evolution of the geography of the same appendage between species is also dependent on *Hox* genes. But for any individual appendage, the evolution of form occurs in the continuing presence of the same Hox protein. A good example is the evolution of insect hindwings.

Wing evolution in insects did not end with the making of the four-winged pattern. In early flying insects, and in the most primitive groups still flying today (such as mayflies and dragonflies), the two pairs of wings appear very similar. In later groups of insects, the two wings have evolved major differences in size, shape, texture, color, and function. In beetles, for example, the hindwing is membranous and used in flight while the hard forewings fold over and protect the delicate hindwings when they are at rest. In butterflies, the shape and color patterns of hindwings are often markedly different from the forewing, such as in swallowtail butterflies. And in mosquitoes and flies, the hindwings are called balancers because they serve as gyroscopes. They are balloon-shaped, and much smaller than the forewing, and sense rotation of the body during flight (figure 7.8).

The evolution of these different hindwing forms are matters of selectively changing the geography of this body part during development. The selective modification of the hindwing is achieved because its development takes place under the command of a specific Hox protein, Ultrabithorax, while the forewing develops without any Hox command. The obvious dependence of hindwing formation on Ultrabithorax is made clear by mutations that abolish Ultrabithorax function in hindwings in beetles, butterflies, and flies—in all cases this causes the hindwing to develop identically to the forewing. All of the details that differ between the respective insect's fore- and hindwings are governed in some way by Ultrabithorax.

The Ultrabithorax protein modifies the regulatory circuitry of beetle, butterfly, fly, and other insect hindwings in ways unique to each group of animals. Ultrabithorax exerts its command by binding to and regulating the activity of switches belonging to wing-pattern genes. This means that in the course of the evolution of different kinds of insects, signature sequences for Ultrabithorax have appeared in the switches of some genes. The coalitions of genes are different in different insects. The genes

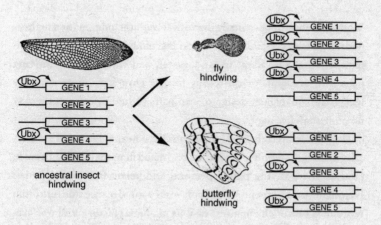

FIG. 7.8 **Evolving hindwing geography**. Different forms of insect hindwings have evolved by changes in the sets of genes controlled by the hindwing-specific Ubx Hox protein. DRAWING BY LEANNE OLDS

that must be shut off to prevent vein formation in the fly hindwing are different from the genes that need to be turned on to make the long tail of a swallowtail butterfly. The geography of hindwings evolves by changing the switches that are governed by Ultrabithorax (figure 7.8).

In a similar fashion, the specialized geography of other individual arthropod appendages also evolves by change in the switches regulated by individual Hox proteins. The long nectar-drinking proboscis of butterflies and the piercing proboscis of mosquitoes, the powerful jumping legs of grasshoppers and crickets, and the claws of crayfish, crabs, and lobster are all products of the selected evolution of individual appendages governed by individual Hox proteins.

The modification of the same limb into many different forms and functions is also the major evolutionary story of our phylum, the vertebrates.

From Fish Fingers to Bat Wings

While vertebrates have made do with fewer appendages then arthropods, they have certainly been no less ingenious in shaping and reshaping their limbs to adapt to land, water, and skies, and in evolving powerful, graceful, or dexterous forms. They have done so while locked into an ancient design of two pairs of limbs bearing generally no more than five digits.

Four-limbed vertebrates—*tetrapods* is the formal term and includes all amphibians, reptiles, dinosaurs, birds, and mammals—first came to land on the evolving paired pectoral and pelvic fins of fish in the Devonian period (about 365 million years ago). Because this transition represents one of the major invasions in animal history and was executed by fairly large animals built of durable skeletons that make good fossils, this has been one of the most intense areas of study in all of paleontology. (It is also the subject of the recent book *At the Water's Edge* by Carl Zimmer, which I recommend highly.)

The topics of central interest to us here are how fins evolved into limbs, and limbs into the great variety of structures we know in tetrapods today. The story can be broken down into four stages. First, we want to understand what existed in fish before tetrapods evolved. Second, we want to know which structures were invented in getting to land. Third, we want to trace how the basic limb design has been changed in evolving different kinds of limbs, such as wings. And fourth, many vertebrates, in adapting to particular niches, have reduced or eliminated limbs altogether, and we will look at a couple of examples, in snakes and fish, of how this was accomplished. In all stages, changing form is a matter of evolving development and I will focus on how the geography of vertebrate limbs has evolved.

The critical time window in the fossil record for understanding the fin to limb transition is in the Late Devonian, about 362–375 million years ago. Fish had already been evolving for 150 million years and a good number of forms with a variety of fin patterns are known from the fossil record. Some of these had long fins that ran the length of the body, some had unpaired dorsal fins, and some had two sets of paired fins that appeared immediately behind a broad head shield. The principal difference between the paired fins of these fish and the limbs of full-fledged tetrapods is the presence in the latter of hands, feet, and digits. Homologs of the upper arm/thigh and lower arm/calf are present in primitive fish fins, but only Late Devonian vertebrates possess the third major limb element, what is called the autopod.

The origin of the autopod is therefore of immense interest. Among the fossil specimens that are most helpful in understanding the transition from fins with two major limb elements to limbs with three are *Sauripteris* and *Acanthostega* (figure 7.9). Both of these animals possessed autopods. In the fish *Sauripteris*, the anatomy of the pectoral fin bears striking similarities to primitive aspects of the tetrapod limb. There are eight, jointed radial bones whose position and number appear to be shared with the digit pattern of primitive tetrapods. These appear to be "fingers" that have evolved in the context of a fin. There

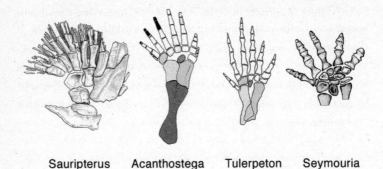

Sauripterus   Acanthostega   Tulerpeton   Seymouria

FIG. 7.9 **From fish fins to fingers.** The early evolution of the tetrapod hand (and foot) as seen in these Devonian fossils involved the reduction in number and specialization of digit form. COURTESY OF NEIL SHUBIN AND MICHAEL COATES, UNIVERSITY OF CHICAGO

are additional similarities in the position and articulation of other fin bones with the tetrapod limb structure. The discovery of a fish with fingers illustrates that intermediate forms do exist in the fossil record, but it does take a combination of skill, patience, and great luck to find the gems (the best *Sauripteris* fossil limb was only discovered in the mid-1990s in a Pennsylvania road cut).

Acanthostega, which appeared a bit later than *Sauripteris*, had four legs, but these could not support the animal's body weight and many features of the limb and body were still fishlike. The front feet bore eight digits, and the rear seven. This early pattern of tetrapod digits is a compelling link with the eight radials of *Sauripteris*. The eight-digit pattern in *Acanthostega* consists of serially repeated structures of no more than five distinct digit types. Later tetrapods would reduce the total number of digits. *Tulerpeton*, a primitive amphibian, had six fingers (figure 7.9) but later tetrapods had no more than five digits, and tetrapods have maintained that upper limit ever since, for more than 300 million years.

I should mention that neither *Sauripteris* nor *Acanthostega* are believed to be the direct ancestors of tetrapods. Rather, these fossil animals represent the sorts of changes that were taking place in several lineages of freshwater fish in the Late Devonian. The similarities in fin and limb anatomy to that of tetrapods are thought to reflect parallel evolution in different groups under similar ecological demands (parallel evolution is an important and frequent phenomenon about which I'll say more shortly).

New Structures via New Switches

How did the autopod evolve? Again, we turn to Evo Devo and the analysis of the genes and embryos of living groups to identify how shifts in limb geography were achieved. The key comparison is between the development of fish fins and tetrapod limbs. In tetrapod limbs, the three elements of the limb are generally specified in a proximal to distal order, with the upper arm or thigh first, the digits last. In fish, the first two phases of limb development are similar to that in tetrapods, but there is no third phase.

In tetrapods, all three phases of limb development involve the deployment of two particular sets of *Hox* genes, a subgroup of two of the four *Hox* clusters. The use of *Hox* genes in the development of the proximal to distal axis of the limb is altogether different from their roles in arthropods (where they generally distinguish one type of limb from another). Between each phase, the spatial patterns of *Hox* expression change and correlate with the specification of each limb element. We know from mutations in *Hox* genes that have occurred in both humans and mice that these *Hox* expression patterns are important for normal limb formation and patterning. Mutations in the *Hox* genes used in the third phase affect the number and size of the digits.

The evolution of the third phase of *Hox* expression in the autopod is a tetrapod invention. This third phase is controlled by separate switches than those controlling the first two phases. It appears that this new structure has evolved because a set of vertebrate *Hox* genes have acquired a new switch or set of switches that activate them in a new distal part of the embryonic limb.

These were not the sole changes involved in the evolution of the autopod. There were many other developmental changes and genes involved in the shaping of the autopod. Other genes, such as members of the bone-promoting *BMP* family and the joint-making *GDF* family, acquired digit-specific switches, and all of the soft tissues—tendons, ligaments, and muscles—and the genes that control their formation and patterning evolved as well.

Flying and Slithering: Evolving Limbs for New Lifestyles

In the ensuing 350 million years, the architecture and function of tetrapod limbs have changed many times in different directions, from spectacular modifications of digit form in evolving wings multiple times, to various degrees of limb reduction in terrestrial and aquatic animals. All of these modifications involve the evolution of limb development, and in several cases Evo Devo researchers have been able to pinpoint some of the important shifts in developing limb geography.

Three separate times—in pterosaurs, birds, and bats—the tetrapod forelimb has been remodeled into a wing for powered flight. To work as a wing, the forelimb must move up and down, forward and backward, and fold against the body when not in use. Interestingly, each of the three vertebrate designs is different in major details. Pat Shipman, in her book *Taking Wing*, has described pterosaur wings as "finger wings," bird wings as "arm wings," and bat wings as "hand wings" (figure 7.10). Let's look at the three designs in evolutionary order, beginning with the pterosaurs.

Pterosaurs took to the air about 225 million years ago, perhaps 70

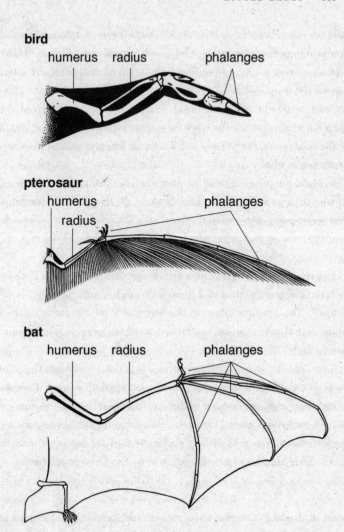

FIG. 7.10 **Wing evolution in vertebrates.** Bird wings are "arm wings" upon which feathers develop along the length of the entire limb. Pterosaur wings are "finger wings"; most of the wing membrane is attached to an elongated digit. The bat wing is a "hand wing" in which the wing surface is attached to multiple digits of the forelimb and extends to the hindlimb. DRAWING BY LEANNE OLDS

million years before birds evolved (birds evolved from feathered dinosaurs, not pterosaurs). The most prominent feature of the pterosaur wing is a very long fourth digit, which supported the outer part of the wing. All parts of the forelimb and digits 1 to 3 were present and the hand bones were fused. Digits 1 to 3 were not part of the membrane attachment. The wing membrane extended along the length of the entire limb, but most of the wingspan was provided by the elongated fourth finger.

In birds, the wings are not membranous, but composed of feathers, which are outgrowths of the skin all along the length of the forelimb. The wingspan is longest in the "forearm" of the limb, and shortest over the bone of the upper arm and the hand and digits. Indeed, the four fingers of birds are very short.

In bats, the wing is made of a membrane that is supported both by the arm and greatly elongated digits 2 through 5, making these "hand wings." The trailing edge of the wing is a membrane stretched between the hindlimbs and is attached at the heel; this aids in stability during flight.

The different architectures of these wings reflect different developmental modifications of a common tetrapod forelimb design. While we know of a very large number of similarities underlying the formation of both bird and mammal forelimbs, the precise differences responsible for the distinct features of bird and bat wings are not yet known in detail. They are under intense study now by Evo Devo researchers.

Some of the developmental shifts are better known for modifications of limbs that have occurred in snakes and a particular species of fish. In snakes, the overall body has been greatly elongated and limb development suppressed. In pythons and boas, a vestigial remnant of the hindlimb still forms, but the forelimb does not. In principle, limb formation could be suppressed at any of several points, from before the initial formation of the limb bud through later stages of its elaboration.

Examination of limb development in python embryos has revealed evolutionary changes in early limb bud formation that account for limblessness. With respect to the forelimbs, the expansion of certain *Hox* gene zones throughout the entire trunk of the animal, extending into the head, has eliminated the site of initiation of the forelimb bud. In the hindlimb, the bud is initiated, but its outgrowth is aborted. The truncation of hindlimb formation correlates with the absence of expression of key signaling proteins, including, for instance, the Sonic hedgehog protein, from organizers in the hindlimb bud. Pythons and boas do develop a small hindlimb spur near the cloaca; more recent families of snakes do not. It is likely that full limblessness in these snakes is due to failure of hindlimb formation at an even earlier stage.

Limb evolution is by no means only a matter of these ancient changes in the origin of major groups of vertebrates. Adaptation and evolution of limb formation continue, and features such as digit number are highly variable in recently evolved species (for example, among salamanders and lizards). In fish, as well, fin evolution is dynamic. One group, the threespine stickleback, has a remarkable recent history that has made it an emerging model for the evolution of vertebrate skeletal anatomy. In many lakes throughout the northern ranges of North America, pairs of stickleback forms occur that have evolved from a common ancestral marine form in very recent history. As the glaciers of the last Ice Age receded beginning some 15,000 years ago, populations of sticklebacks were isolated in glacial lakes. Then, in a geologically brief interval, these populations have evolved into forms that occupy different niches: a shallow-water, bottom-dwelling, short-spined form and an open-water, long-spined form (figure 7.11).

These forms differ particularly in their body armor, which includes hard plates on the sides of the body, and spines projecting from the upper and lower body. Spine number and length are related to predation pressure. In the open water, longer spines help protect the sticklebacks from the gape of predators. But, on the bottom, long pelvic

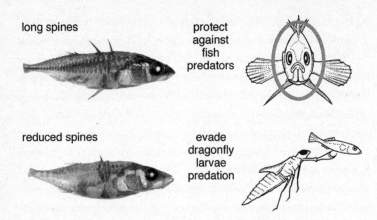

FIG. 7.11 **Evolution of spine number in stickleback fish.**
Long spines protect open-water forms from the gape of predators
by making the diameter of the fish larger. But, long spines are a
liability in bottom-dwelling forms that are fed on by dragonfly
larvae, so the reduction of spines helps to reduce this predation.
COURTESY OF DAVID KINGSLEY, HOWARD HUGHES MEDICAL INSTITUTE
AND STANFORD UNIVERSITY

spines can be a liability. Amazing as it may sound, dragonfly larvae are
voracious predators and can actually grab the sticklebacks by their
spines. Under this strong pressure, spine-deficient forms have evolved
repeatedly in natural populations.

The pelvic spines are part of the fish hindlimb, so their reduction is
a modification of the development of the hindlimb skeleton.
Developmental biologists have learned a good deal about a number of
genes involved in the formation and differentiation of forelimbs and
hindlimbs. One of these genes, called *Pitx1*, is involved in making
hindlimbs in tetrapods and the pelvic fin in fish. Analysis of *Pitx1*
expression in spine-deficient sticklebacks from a lake in British
Columbia reveals that expression is specifically lacking in the pelvic fin

buds of these fish. This evolutionary change in *Pitx1* regulation appears to be caused by, you probably guessed it, a change in a switch of the *Pitx1* gene that abolishes its expression in the hindlimb. The evolutionary change in this switch has allowed *Pitx1* function to change in the pelvic fin, without altering the gene's essential functions elsewhere in the developing fish.

Moreover, analysis of a second population of spine-reduced sticklebacks from Iceland suggests that the same change in *Pitx1* expression has occurred independently. Spine reduction is also evident in fossil species, and in distantly related fish genera. This suggests that pelvic spine reduction is a frequent, independent occurrence in the evolution of some fish, and may repeatedly involve evolutionary changes in *Pitx1* gene switches. These observations, and the example of maxilliped evolution described in the last chapter, show that some evolutionary changes are not rare, once-in-history occurrences, but that under similar selection pressures, similar changes occur in different populations and species. In this sense, evolution is "reproducible."

Indeed, fin or limb reduction is not at all rare. Two separate groups of mammals—cetaceans (whales and dolphins) and manatees—have greatly reduced their hindlimbs as they independently evolved from land-dwelling ancestors into their fully aquatic lifestyles. In addition, legless lizards have evolved repeatedly. The sticklebacks are therefore not some obscure oddity, but important models of common, yet profound evolutionary transitions. Furthermore, their recent fossil record is remarkably well preserved in certain locales in deposits that encompass many millennia. This record indicates that pelvic reduction can evolve in fewer than 10,000 generations, or under 10,000 years. While this is certainly not instantaneous in terms of real time, it is a very brief interval of geologic time. The combination of their exquisite fossil record, the detailed genetics of living populations, and the repeated independent examples of similar evolutionary changes makes the sticklebacks one of the most compelling case studies of evolution.

Four Secrets of Evolutionary Innovation

The animals and structures highlighted in this chapter reveal several of the most important secrets to the evolution of new forms. I will briefly reiterate and expand on four of these. The first secret of evolutionary innovation is, without a doubt, to work with what is already present. It is perhaps worth pondering what didn't happen in these animals. In spiders spinnerets did not arise *de novo* on the body, nor in vertebrates did wings sprout from the top or sides of a four-legged animal. Rather, these and all of the other structures are modifications of preexisting limbs. More than twenty-five years ago, François Jacob captured this important essence of evolution in an essay entitled "Evolution and Tinkering." Jacob pointed out that nature works more like a tinkerer, by working with and cobbling together available materials and constantly modifying and retouching structures over eons, and not as an engineer would with a preconceived plan and specialized tools. This message extends to the gene level, in that we find the same "old" genes being used in different ways. The most common path to evolutionary innovation is a zigzagging course from A to B, not straight to B from scratch.

The second and third secrets, multifunctionality and redundancy, were first recognized by Darwin. I have emphasized the opportunities presented when these two dimensions are present. Any part of a multifunctional structure that is at least particularly redundant in function sets the stage for specialization through the division of labor into two structures.

The fourth secret to innovation is modularity. Way back in chapter 1, I said that I believed that the modular architecture of arthropods and vertebrates was in part responsible for their great success. Look at what the modularity of arthropods has accomplished—the evolution of many different adaptations that co-occur on the same animal, and inventions that have given rise to the most diverse group of animals on Earth. In vertebrates, the ability to evolve a long fourth finger in pterosaurs, or many long fingers in bats upon which to extend a wing membrane, or to evolve hundreds of vertebrae in snakes to lengthen the body, or in stickle-backs to selectively reduce pelvic/hindlimb structures, is owed to each animal's modular design. Modularity allows for the modification and specialization of individual body parts, sometimes in the extreme, inde-pendent of other body parts.

Underlying the anatomical modularity of the adult animals is a modular embryo geography, and the modular genetic logic of switches. These switches allow evolutionary change to occur in one part of structure, independent of other parts. Switches are the secret to modu-larity, and modularity the secret to arthropod and vertebrate success.

The consequences for biodiversity should now be self-evident. Innovation allows for invasion of new niches, and invasion leads to the expansion of diversity.

To this point, I have concentrated almost entirely on big shifts in body and limb design, those differences that define the higher groups of arthropods and vertebrates. It is fine to speak of "birds," "bats," and "beetles" in a general sense, but keep in mind that these names actually represent broad categories containing many species— hundreds of bats, thousands of birds, and hundreds of thousands of beetles. The success of each group is owed in part to the major innova-tions I have described, but the abundance of each kind of animal is also a product of expanding into many niches, and often of additional innovations (e.g., sonar in bats, webbing in waterfowl, elaborate songs for communications). In the next chapter, I am going to zoom in on just one group, the butterflies, to illustrate how one innovation, the wing, has provided the foundation for subsequent rounds of innova-tion and explosive diversification.

Drawings and notes from the notebooks of explorer-naturalist
Henry Walter Bates. COMPOSED BY JOSH KLAISS

8

How the Butterfly Got Its Spots

"Evolution is chance caught on the wing."

—Stuart Kauffman, *At Home in the Universe*,
paraphrasing a translation of Jacques Monod
in *Chance and Necessity*

AFTER ELEVEN YEARS in the Amazon, having collected 14,712 different animal species (8000 of which were new to science), his body wracked by tropical disease, poor nutrition, and prolonged exposure to sun and heat, and having endured robbery, abandonment by servants, and other deprivations, Henry Walter Bates left the jungle for

England in June 1859. His timing was fortunate—in just a few months Darwin's *The Origin of Species* would appear.

The two voyagers would fast become great friends. Bates was an immediate adherent to Darwin's views and he initiated a correspondence that would last more than twenty years, right up until Darwin's death. Bates was excited and convinced that his observations and collections would lend support to Darwin's theories. "I think I have got a glimpse into the laboratory where Nature manufactures her new species," he wrote in one of his first letters to Darwin.

Bates's greatest contribution was his discovery of what he called "analogical resemblance," or mimicry. Bates had studied many cases in insects, especially butterflies, where protection was afforded by one species assuming the coloration of another. He noted that birds found certain butterflies to be edible and others noxious. Birds learned to distinguish types based upon only a few experiences. Bates observed that some palatable butterflies mimic the color patterns of unpalatable forms that birds learned to avoid, which affords them protection from predation. This illustration of natural selection in butterflies thrilled Darwin, who told Bates that his paper describing mimicry was "one of the most remarkable and admirable papers I have ever read in my life." The phenomenon is still known today as Batesian mimicry (figure 8.1).

Darwin continually milked Bates for insights from his vast and superb original knowledge of natural history, particularly for Darwin's ongoing (and now famous) work on sexual differences and selection. Bates received great encouragement from Darwin, especially to write and publish a narrative of his travels. Not only did Bates draw upon Darwin's views, but Darwin even reviewed, edited, and wrote an "appreciation" for the one book Bates produced in his entire career, *Naturalist on the River Amazons* (1863). Darwin had predicted it to be a great success and he was right, for Bates's writing proved to be superior to either Darwin or his original companion in the Amazon, Alfred Russel Wallace. Bates's book is still a terrific read today.

Among those 14,000 plus species collected were many butterflies,

Amauris niavius dominicanus

Danaus chrysippus

Amauris albimaculata

P. dardanus f. hippocoonides (mimic)

P. dardanus f. trophonius

P. dardanus f. cenea (mimic)

FIG. 8.1 **Batesian mimicry.** All of the butterflies on the top row are distasteful to birds. Each variation of the swallowtail *Papilio dardanus* on the bottom row is a mimic of the form above. Note the extensive similarity among each pair of entirely different species.
PHOTOS COURTESY OF DR. PAUL BRAKEFIELD, UNIVERSITY OF LEIDEN

over 550 species from the region of Ega alone. Bates saw the value of his treasures through a Darwinian lens: "no descriptions can convey an adequate notion of the beauty and diversity in form and colour of this class of insects in the neighborhood of Ega. I paid special attention to them, having found that this tribe was better adapted than almost any other group of animals or plants, to furnish facts in illustration of the modifications which all species undergo in nature under changed local conditions."

Bates continued with my favorite passage: "It may be said, therefore, that on these expanded membranes nature writes, as on a tablet, the story of the modifications of species, so truly do all changes of the organization register themselves thereon."

He concluded: "Moreover, the same colour-patterns of the wings generally show, with great regularity, the degrees of blood-relationship of the species. As the laws of nature must be the same for all beings, the conclusions furnished by this group of insects must be applicable to the whole organic world; therefore the study of butterflies— creatures selected as the types of airiness and frivolity—instead of being despised, will someday be valued as one of the most important branches of Biological science."

Bates's passion for, and convictions about, the value of butterflies to science have been shared by many naturalists, professional and ama- teur, since he wrote these words 140 years ago. The stories on the expanded membranes of butterfly wings not only delighted Darwin, but William Bateson also paid particular attention in his great book to individuals with atypical patterns. Since then, naturalists have charac- terized other types of mimicry in butterflies (of other butterfly pat- terns, of owl eyes, dead leaves, or even bird droppings), and this group of animals has inspired many evolutionary and ecological studies. Of course, the fascination with butterflies extends far beyond conven- tional scientists. The novelist Vladimir Nabokov had a lifelong passion for butterflies. His expertise supported him, as a curator of Lepidoptera at the Harvard Museum of Comparative Zoology, before his writing achieved wide acclaim.

In this chapter, I will explore the wonderful world of butterfly wing patterns. In this group of insects, the wing has served as a canvas for the evolution of thousands of color patterns. I will focus on the inven- tion of color-patterning systems, and on how so many variations have evolved. We will see that butterflies provide the most exquisite exam- ples of how new patterns evolve when very old genes learn new tricks.

Making Sense of Wing Patterns

I did not have to travel halfway around the world nor endure any of the hardships Bates did in order to begin to learn about butterflies. My journey began in a parking lot on the Duke University campus in

Durham, North Carolina. I was there many years ago to give a seminar on my laboratory's work, which at the time focused on how genes govern the number and position of bristles on the body of a fruit fly. As is the academic custom when lecturers visit other universities, I was scheduled to meet with a number of Duke's biology faculty. But one of the professors I was to meet was running late because a pipe had burst in his home. I almost didn't get the chance to meet Fred Nijhout that day and, had I missed him, not only would this chapter not exist, but I would have missed some of the most thrilling moments I have ever enjoyed in the laboratory.

The bristle patterns of flies has been a great model for understanding some general mysteries in development, such as how structures are located at precise positions on the body. However, in the rush of the few minutes we had crossing that parking lot to get to my next meeting, Fred asked me whether the rules we were finding out for fruit fly bristles could explain his primary love—the patterns on butterfly wings. Frankly, I had no idea.

Whenever I had looked at butterfly wings, accustomed as I was to the pale wings of fruit flies, I just saw chaos. Psychedelic patterns and colors running in all directions. Lines, spots, squiggles, blotches—I couldn't make out any order (I have the same grasp of modern art). But Fred's question haunted me for many months. I knew the fantastic lore of butterfly mimicry, predator avoidance, and sexual selection. This was a gold mine, if one could just get some kind of grip on those patterns and the genetic and developmental mechanisms for making them.

Fortunately, Fred soon published a book that served as a primer on all matters of butterfly biology. I learned that one could break down the chaos of those patterns into some order. In the 1920s and 1930s, some comparative biologists perceived an overall plan to butterfly wing patterns. This "ground plan" represented an idealized picture, from which individual species diverged to different degrees. The ground plan consists of several pattern elements near the base of the wing, in the central part, and toward the edge of the wing, which are repeated in each subdivision of the wing that is bounded by veins. These pattern elements consist of sets of bands of different width, as well as the eyespots (figure 8.2). The subdivisions of the wing are serial homologs, and the patterns within these subdivisions are therefore modular.

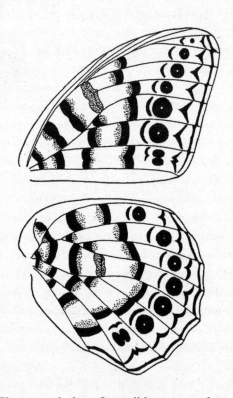

FIG. 8.2 **The ground plan of possible pattern elements.** This schematic represents the idealized spectrum of all possible elements in Nymphalid butterflies. Note the serial repetition of pattern between adjacent wing subdivisions. COURTESY OF DR. H. FREDERIK NIJHOUT, FROM HIS *THE DEVELOPMENT AND EVOLUTION OF BUTTERFLY WING PATTERNS*; USED BY PERMISSION OF SMITHSONIAN INSTITUTION PRESS

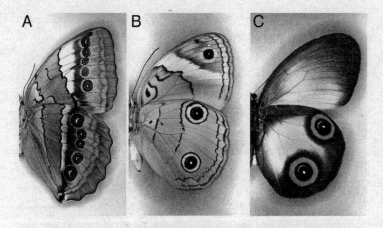

FIG. 8.3 **Variation on the ground plan.** The three species, *Stichophthalma camadeva* (A), *Fauris menado* (B), and *Taenaris macrops* (C), exhibit different degrees of representation of the ground plan, from virtually all elements to very few. PHOTO COURTESY OF DR. H. FREDERIK NIJHOUT

Butterfly wing patterns are generally composed of some subset of the maximum ground-plan pattern, ranging from species such as *Stichophthalma camadeva*, which displays most of the ground-plan elements, to those with just a few (figure 8.3). A survey of the thousands of living butterfly wing patterns reveals that diversity is largely a matter of loss of particular elements, or the modification and repositioning of these elements. Patterns that appear more chaotic are due to the dislocation and misalignment of bands between adjacent subdivisions of the wing.

The most crucial observation about these patterns is that each individual band or spot appears to be able to evolve its shape, color, or size independently of other elements. This indicates that the development of individual pattern elements can be uncoupled from one another.

What Did Butterflies Invent?

The spectacular beauty and diversity of butterfly wings are due to at least three inventions that took place after this lineage split off from other insects. These inventions include wing scales, coloration, and geometrical patterning systems.

Scales are the fundamental units of butterfly and moth wing patterns. The order to which these bugs belong, Lepidoptera, gets its name from the Greek *lepis*, meaning scale or flake, and *ptera*, meaning winged creature. Scales were invented before the elaborate wing-color patterns and probably initially served quite a practical use. If you have ever grabbed a moth in your hand or between your fingers, you have noticed the "dust" residue—these are scales. The easy detachment of these scales is an advantage for these large-winged animals in freeing themselves from sticky places, such as spiderwebs.

Moth and butterfly wings are entirely covered with scales, each of which is the product of a single cell (figure 8.4). Entomologists have believed for a long time that scales evolved as modifications of sensory bristles, becoming flat and wide instead of long and slender, and losing their sensory innervation. This scenario has been borne out by Evo Devo studies. Ron Galant in my laboratory found that developing scales use one of the tool kit genes that is also used for making bristles in the fly, suggesting that scales are indeed modified bristles.

The colors found in butterflies are rivaled by few other insects. Each individual scale is a particular color, which can be seen at high magnification, where individual scales may be an entirely different hue than their neighbors (plate 8a). Any perception of blending or intermediate tones is a visual effect of the spatial arrangement of individual scales of discrete colors. The colors in the wings are due to both chemical pigments and structural colors. Iridescent blues and greens, as well as powdery whites, are the result of the way scales absorb, reflect, and scatter light. Different structural colors are due to very fine differences

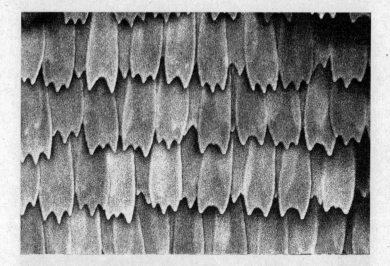

FIG. 8.4 Butterfly wing scales. STEVE PADDOCK

in scale microanatomy, as well as the combined effects of scale structure and the making of certain chemical pigments.

The geometric patterns on wings are due to the invention and elaboration of developmental pathways that organize them. We know the most about one kind of pattern element, the eyespots. These spots are composed of concentric rings of scales of different colors (plate 8b). Many studies have investigated their role in helping butterflies to evade predators. The proposed role of the eyespots in many species is to deflect the attention of attacking predators (usually birds or lizards) toward the outer edge of the wing, away from the vulnerable main body. Butterflies with even good-size pie slices taken out of their wings can still fly (figure 8.5), whereas a blow to the body can be fatal. The eyespots may draw attention, either because of their conspicuous markings that stand out in contrast to the rest of the wing body or since their resemblance to eyes provokes the predators' instinct to attack these patterns.

The critical role of eyespots in butterfly defense and their tremen-

FIG. 8.5 **Predator damage on a butterfly.** This *Bicyclus anynana* butterfly has been attacked, but because the damage is localized to the edge of the wing, it can still fly and reproduce. PHOTO (TAKEN IN KENYA) COURTESY OF PAUL BRAKEFIELD, UNIVERSITY OF LEIDEN

dous diversity among species prompted us to concentrate our efforts on understanding how these pattern elements are made and evolve.

Making Eyespots: Teaching Old Genes New Tricks

The adult wing patterns we see are the realization of a process that begins in the caterpillar. Each wing forms from a flat disc of cells that grows enormously during the many stages of larval development (most butterflies have five larval stages). The caterpillar then forms the chrysalis and it is in the chrysalis, just before the butterfly emerges, that the final color pattern is filled in. Though invisible to the naked eye, some parts of the future wing pattern are being made in the caterpillar when the wing is a tiny immature disc, just a fraction of its adult size. This is about a week or more before the adult butterfly will emerge from the chrysalis. In chapter 2, I described one of the most spectacular experiments that revealed early events in butterfly wings, those of Fred Nijhout on transplanting the eyespot focus. Fred showed that the position of the future eyespots is decided in the caterpillar. He discovered that the concentric ring patterns of eyespots are induced by an organizer, the so-called focus, that lies at the center of the developing eyespot.

Because Fred's transplantation experiments revealed the presence of a novel organizer in the developing butterfly wing, we set about in my laboratory to try to identify genes involved in making the eyespot. Our main questions were: What sort of genetic system generates individual patterns? How did this system evolve? What are some of the genetic tools used to inscribe Bates's "tablets"? Did the butterflies evolve new genes for making spots, or did they use what was already available?

We had some hunches to go on. My laboratory and several others had made pretty good progress in identifying tool kit proteins involved in making a fruit fly wing. Our logic in tackling the making of butterfly wings was to rely on the evolutionary relationships among insects. Since insect wings evolved just once, then what we knew about the making of fruit fly wings should apply, in general, to the making of butterfly wings. Our hope was, if we were lucky, that studying the butterfly counterparts of the fruit fly tool kit might lead us to clues about the unique features of butterfly wings.

We were lucky.

A team of scientists in my lab isolated from the Buckeye butterfly a number of tool kit genes whose homologs were known to be involved in the building and patterning of fruit fly wings. The presence of these genes in butterflies wasn't a surprise, nor would that constitute evidence for their role in patterning wings. The key was to see if we could place any of these genes in the scheme of butterfly wing patterning, at the time when transplantation experiments told us that patterns were being set up. To do our experiments, we had to look at where genes were expressed in the tiny wing discs in caterpillars. We wanted to see, through a microscope, hints of how the beautiful patterns that would appear later in the adult were made.

We found that all of the butterfly genes were expressed in parts of the developing butterfly wing disc that corresponded to the same geographical regions where they are deployed in fruit fly wings. This told us that there is a common geography to developing insect wings. The top and bottom surfaces of the butterfly wings, the front and rear of each wing, and the edges of the wing are staked out by the same genes in both species. This is a nice affirmation of the conservation of an ancient wing design. But more intriguing, and exciting, were patterns of gene expression we saw in butterfly wings that had no counterpart in fruit fly wings. I will never forget the moment when my technician Julie Gates called me to the microscope to see the most stunning pattern, of beautiful spots, in the caterpillar wing discs. We saw two pairs of spots on each disc precisely where the eyespots would appear a week later in development, in the positions that Fred Nijhout had defined as the foci of eyespots (plate 8c). Fantastic.

The spots were made by just one of the dozen or so genes we studied. You have already read a lot about this gene—it was *Distal-less*. This was tremendously exciting because it meant that the same gene involved in building fruit fly limbs and arthropod limbs appeared to be doing something altogether new in butterfly wings. *Distal-less* still kept its old job: it was also deployed in the distal parts of all developing butterfly limbs, just as in all other insects and arthropods. The spots of *Distal-less* expression in butterfly wings were a new trick, "learned" long after its ancient role in limb-building (figure 8.6). Remember, everything about a tool kit protein's action depends on context. *Distal-*

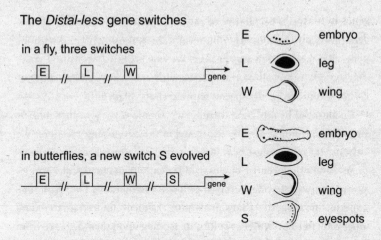

The *Distal-less* gene switches

in a fly, three switches

E	leg... embryo
L	leg
W	wing

in butterflies, a new switch S evolved

E	embryo
L	leg
W	wing
S	eyespots

Fig. 8.6 **A new genetic switch for eyespot expression evolved in the butterfly *Distal-less* gene.** Switches in the *Distal-less* gene control expression in the embryo, larval legs, and wing in both flies and butterflies, but butterflies have evolved an additional switch to control expression in eyespots. drawing by leanne olds

less carries out its limb-building role in specific places and times; its job in the wing spots is in another place and time, and controls an altogether different pattern.

How did *Distal-less* learn the new trick of making spots in the wing? The gene acquired a new switch that responded to the specific longitude and latitude coordinates of these spots of cells. *Distal-less* spots always form exactly between two veins and along the outer edge of the wing. The precise and reproducible coordinates of these spots tell us that there are tool kit proteins active at these positions that flip on the switch in the *Distal-less* gene.

The discovery of *Distal-less* expression in the developing eyespots of the Buckeye butterfly gave us the foothold we were hoping for. It told us that we would be able to make sense out of how these intricate patterns were made. One of the first worries we had to confront was whether we had discovered something general about butterfly wing patterns, or just a peculiarity of this one species. So we looked at how *Distal-less* was used in other butterflies, both those with and without spots. The correlation was perfect. We saw beautiful spots of *Distal-less* expression in all of the species with spots; we saw no spots of *Distal-less* in any of the species without spots (plate 8d).

Encouraged by our good fortune, we went looking for other tool kit proteins that might also be expressed in the developing eyespots. We believed that there must be other proteins to discover because eyespots were made of concentric circles of differently pigmented scales; somehow, each ring of scales received different instructions. Fred Nijhout's experiments suggested that signals from the focus induced surrounding rings of cells to be different colors, depending upon their distance from the focus. *Distal-less* marked the cells at the center, but the outer rings must also be marked somehow.

We got lucky again. In my lab, postdoctoral fellow Craig Brunetti was searching for other tool kit proteins in butterfly eyespots and he found two more spectacular patterns. When he looked very carefully at the expression of two tool kit proteins called Spalt and Engrailed, he saw that they were expressed in a spot and a ring, respectively, in the African species *Bicyclus anynana* (plate 8e). The eyespots in this species have a white center, surrounded by a wide black ring, which is surrounded in turn by a gold ring. The Spalt pattern precisely marked the future black ring; the Engrailed pattern did the same for the gold ring. (The detailed correspondence between the rings of protein expression in developing scales and the future rings of the eyespots is exquisite in the color plate.)

Engrailed and *Spalt* are also very old genes with other jobs, so the explanation for their new role in butterfly eyespot patterning is the same as for *Distal-less*: the evolution of new switches controlling each gene has enabled these genes to add a new job in butterflies.

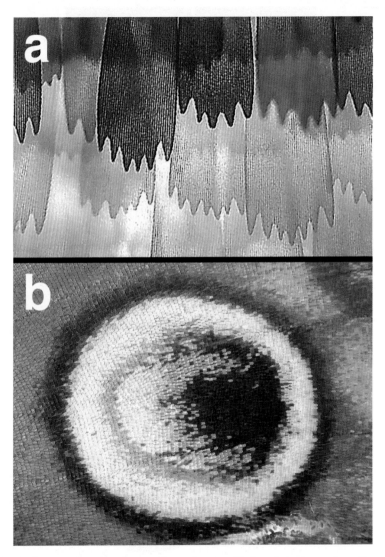

8a. A close-up view of scales on a butterfly wing. Each scale is the product of a single cell.

8b. A close-up view of a butterfly eyespot reveals the color pattern formed by rows of scales. Note that each scale is one color, and "stray" scales exist in fields of other colored scales.

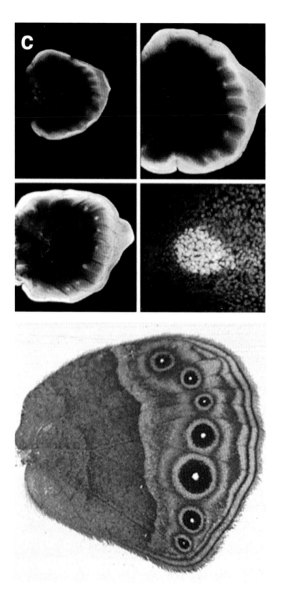

8c. The expression of the tool kit protein Distal-less in a caterpillar wing marks the future centers of butterfly eyespots. The Distal-less protein (revealed in green) resolves into small spots (visible at high resolution in the lower-right-hand panel of the smaller photos). The spots mark the position of the white centers (larger photo) of the adult eyespots that will develop more than a week later.

8d. The expression of the Distal-less protein in spots correlates with the presence or absence of spots in different species. In the left column, Distal-less expression in developing wings of caterpillars of four species; the corresponding adult patterns are shown at right.

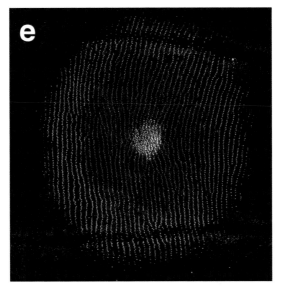

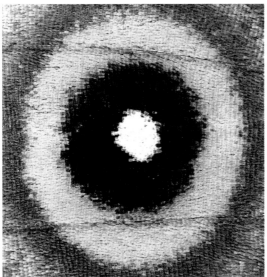

8e. Different future rings of the adult eyespot are marked by the expression of two different tool kit proteins in the chrysalis stage (in green and purple; they overlap at center). These rings mark the future white, black, and gold rings of the adult that will form more than a week later. CRAIG BRUNETTI

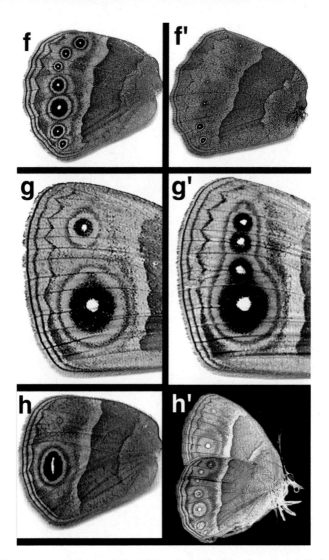

8f. Different seasonal forms of the African butterfly *Bicyclus anynana*. Left, hindwings of wet season form; right, hindwing of dry season form with tiny eyespots. COURTESY OF PAUL BRAKEFIELD, UNIVERSITY OF LEIDEN

8g. The *Spotty* mutant makes two extra eyespots on its forewing. COURTESY OF PAUL BRAKEFIELD, UNIVERSITY OF LEIDEN

8h. Mutant butterflies with large spots, irregular spots, or altered colored spots. COURTESY OF PAUL BRAKEFIELD, UNIVERSITY OF LEIDEN

8i. Selection for large and small eyespot size in *Bicyclus anynana*. Top row, dry and wet season forms. Middle row, small-eyespot-selected line grown at dry or wet season temperatures. Bottom row, large-eyespot-selected line grown at dry and wet season temperatures. COURTESY OF PAUL BRAKEFIELD, UNIVERSITY OF LEIDEN

8j. Two forms of tiger swallowtail butterfly. The melanic female is a mimic of the pipevine swallowtail butterfly.

8k. Mimicry in *Heliconius* butterflies. Red and yellow are warning colors in butterflies. Each row depicts *H. melpomene* (left) and *H. erato* (right) specimens found in same geographic area. Note the geographic differences within species and the remarkable resemblance among individuals of different species found in the same area. COURTESY OF H. FREDERIK NIJHOUT, FROM HIS *THE DEVELOPMENT AND EVOLUTION OF BUTTERFLY WING PATTERNS*; USED BY PERMISSION OF SMITHSONIAN INSTITUTION PRESS

My Fifteen Minutes

Let me briefly interrupt the science here with a little story about the unexpected consequences of unexpected discoveries. It taught me a big lesson about the importance of beauty to general interest in science.

Well after the thrill of first seeing *Distal-less* in spots, we wrote up our results for formal publication. We quickly found out that not everyone would share our excitement. The journal *Nature* rejected the paper without further review. Ouch. But, if at first you don't succeed . . . So, we sent the article to the journal *Science*. Its editors were much more receptive and decided to publish our work, and to run a picture of a butterfly wing on the cover of that issue. This was just great for the team—mission accomplished.

However, in other ways, the story was only just beginning.

It usually takes at least a couple of months for a paper to appear, so I had put that article out of my mind when I was later away at a scientific conference. I was staying on a university campus in a dorm room, eating cafeteria food and indulging in the usual glamour of academic science. In between listening to colleagues' presentations, I received a message to call Nicholas Wade, a science writer for *The New York Times*.

Baffled, I gave him a ring and discovered that he was preparing a feature story on our article, which was about to appear in *Science*. I thought this could be neat, something for Mom and my neighbors to read to see how I have used all those years of college and nights in the lab. We had a long discussion and I returned to the meeting and more days of talks.

A little media attention goes a long way. The *Times* feature prompted many other papers to cover the story. This was summertime, and one prominent newspaper told me they wanted something nice to put on the front page to push the then ongoing, infamous O. J. Simpson murder trial off of page one. So various papers wrote articles about finding "the secret of beauty," as one story termed it.

Then came TV. I was sitting down to dinner when I recognized a picture of ours being shown on a national news program. Astonished, I kept watching and this was followed by a long video essay by Roger Rosenblatt, who had been prompted by our article to contemplate the question of whether our sense of wonder and beauty is diminished or enhanced by scientific understanding (I think you can guess my opinion).

Months later, *Time* magazine decided that I should be recognized among a group of young Americans. I wound up in a tuxedo at a dinner with the President, the Washington press corps, and various movie stars and politicians (footnote: a lot of actors are shorter in real life than they appear on screen).

The craziness had one more chapter. Out of the blue, I got a call from a top Hollywood producer who saw the *Time* article and just wanted to have a chat, in person. Of course, I went to L.A. and we had a great time talking about science, movies, and butterflies.

Okay, so now I get it. Butterflies do inspire broad interest, and I am grateful for that, and my fifteen minutes in the limelight. I still get a lot of good-natured ribbing from colleagues about this whole episode.

Of course, there are critics everywhere, so I can't resist sharing one piece of anonymous mail I received during my little media circus (figure 8.7). I can't wait until this guy hears about this book.

How Butterflies Change Their Spots

In Kipling's fairy tale, once the leopard got his spots, he was quite contented and would never change again. But butterflies are of a different mind than the leopard, and their evolution has involved changing their spots many, many times. This is obvious when looking at different species, but I will begin with the tale of a butterfly that routinely changes his spots with the changing of the seasons in Malawi.

All I have learned about *Bicyclus anynana* has been taught to me by Paul Brakefield and his students at the University of Leiden in the

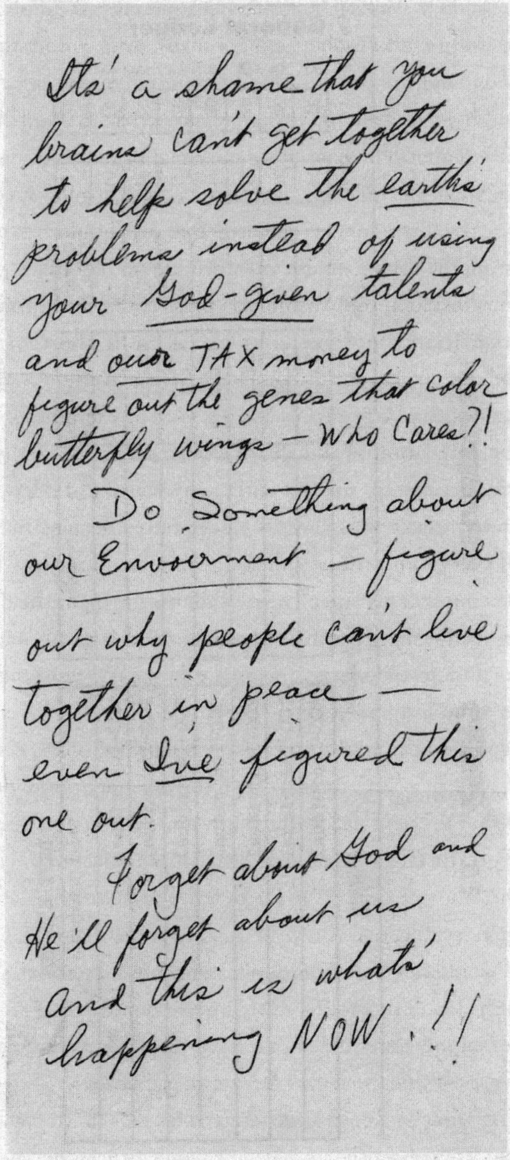

Its' a shame that you brains can't get together to help solve the earth's problems instead of using your God-given talents and ouor TAX money to figure out the genes that color butterfly wings — Who Cares?!

Do Something about our Envoirment — figure out why people can't live together in peace — even Ive figured this one out.

Forget about God and He'll forget about us And this is whats' happening NOW.!!

Fig. 8.7 Fan mail.

Netherlands, and by Vernon French at the University of Edinburgh. Paul has studied this remarkable butterfly for many years in both the field, in Malawi, and in huge populations he maintains in his lab in Leiden.

In the wilds of Malawi, *B. anynana* has adapted to the pronounced seasonal differences in its habitat by learning to change its spots. In the wet season, when the foliage is green and abundant, the species bears large conspicuous eyespots on its wings that aid it in surviving attacks from birds and lizards (plate 8f, left). But in the dry season, when the foliage has withered and the leaf litter is brown, and the butterfly must be less active, those large eyespots serve as bull's-eyes that make the wings stand out against the brown backdrop—screaming, "Here I am, eat me!" So as the weather cools and gets drier at the end of the rainy season, the last broods of caterpillars and chrysalides sense the change, and they emerge instead with no eyespots, just a few tiny little flecks of color where the eyespots would have been (plate 8f, right). These dull brown butterflies will rest hidden on dead leaf litter and wait out the long dry season before the rains return, then they will mate. Their offspring, growing in a warm, humid environment, sense the climate and develop large eyespots that will help protect them in their much more active state.

The adaptation of these butterflies is no "just so" story. Paul and his students have released butterflies with big eyespots in the dry season and found that they are picked off much more frequently than the dull brown, cryptic form, so the evidence for natural selection is clear in the wild. In the laboratory, rearing the developing butterflies at different temperatures reproduces the wild patterns—temperatures of 73° F. (about 23° C.) produce the wet season form; at 62° F. (about 17° C.) the dry season form develops. By shifting the temperature at different stages, Paul's group determined that the critical period that determined the size of the eyespots was the late caterpillar stage.

When my student David Keys studied Distal-less expression in *B. anynana* caterpillars raised at different temperatures, he saw an exact correspondence between temperature, the number of cells expressing the Distal-less protein, and the size of the adult eyespots. At low temperatures, fewer cells expressed Distal-less in the spots, while at higher temperatures the number of Distal-less expressing cells was much greater. In this species, the eyespot switch in the *Distal-less* gene responds differently at different temperatures. We do not think that the switch itself senses the temperature directly, but that levels of certain hormones produced elsewhere in the caterpillar's body vary depending upon the season and temperature. Hormones in insects, just like hormones in our bodies, regulate stages of development and the development of certain tissues. The effects of hormones are ultimately mediated through genetic switches. The *Distal-less* wing spot switch has evolved a hormone-responsive signature sequence in *B. anynana* that enables it to respond to environmental changes.

The capacity to control spot development in response to seasonal changes is just one example, albeit a vivid one, of how development and form evolve under natural selection. In the course of butterfly evolution, all sorts of spot patterns have arisen. For example, in the genus *Bicyclus* alone there are eighty species, which differ from one another in the size, position, and, occasionally, their number of eyespots. This suggests that it is fairly "easy" for butterflies to evolve new wing patterns. There may be a greater degree of freedom in the evolution of possible butterfly wing patterns than in the evolution of other structures. The reason for this flexibility may be that the genetic regulation of wing patterning is organized so that mutations can occur that affect only wing patterns but do not affect other body parts. Evolution in butterflies is very much a matter of "chance caught on the wing."

We can get a window into the way butterfly wing patterns evolve from the variation in wing patterns seen in the lab and in the wild. Paul Brakefield and his colleagues have isolated a number of very striking, spontaneous mutants with different eyespot patterns. Some of these mutants display no other changes in body pattern. They could represent a class of mutations that, because of the restriction of their effect to the wings, might be viable variations in the wild. Indeed, one of these mutants, named *Spotty*, displays four eyespots on the forewing instead of the usual two (plate 8g, right). In a closely related species, *B. safitza*, variants with four eyespots have often been noted in the wild. It is pretty easy then to imagine how the number of eyespots could evolve in this group. Similarly, Paul has isolated mutants that change the color scheme, size, or shape of eyespots (plate 8h) and that thus resemble the sorts of differences that exist among closely related species.

Another angle on the process of wing evolution is the simulation of natural selection by breeding experiments in the lab. In these studies, Paul and his students, instead of birds and lizards, determined the fate of butterflies with spots of different sizes. In any population of butterflies, there are slight variations in the size of the spots. This can be the raw material for natural selection in the wild or "artificial selection" in the lab. Paul and his team established two distinct populations of butterflies, one by selecting and mating butterflies with the largest spots reared at cooler temperatures, and a second line by selecting and mating butterflies with the smallest eyespots reared at warmer temperatures. After about twenty generations of this artificial selection scheme, they obtained populations of butterflies that either made large or small eyespots, independent of temperature (plate 8i).

What happened in these experiments was that the existing variation in eyespot size, which is due to genetic variation in the starting populations of butterflies, was selected for at two extremes (large and small). This resulted in morphologically and genetically different populations. This is, in essence, just what happens in the wild, but over time periods typically much longer than twenty generations.

These explorations into the changeability of eyespots in *Bicyclus* reveal some of the possibilities in butterfly wing-color pattern evolution, a spectrum that in reality butterflies have done a remarkable job in exploring. Among other eyespot-bearing species, the numbers, sizes, and color schemes of wing eyespots have evolved a fantastic degree of diversity. Underlying the diversity of butterfly wing patterns must be different developmental instructions. The discovery of the Distal-less,

Engrailed, and Spalt proteins expression in eyespots has given us ways to see how the many variations on the eyespot theme have evolved in different species.

The most obvious difference between species is the evolution of eyespot number. The evolution of the number of Distal-less spots in wing discs exactly tracks the evolution of eyespot number. This tells us that evolutionary changes in *Distal-less* regulation have evolved between species, and it shows how one innovation, the evolution of spots, leads to further diversity in patterns. Once Distal-less-expressing eyespots evolved, tinkering with Distal-less expression produced butterflies with fewer or more eyespots, different-sized spots or, as in *B. anynana*, seasonal changes in eyespots. These changes in *Distal-less* regulation were most likely accomplished by changing the signature sequences of the *Distal-less* gene eyespot switch (figure 8.8).

The *Distal-less* switch ⟨S⟩ has been modified in several ways

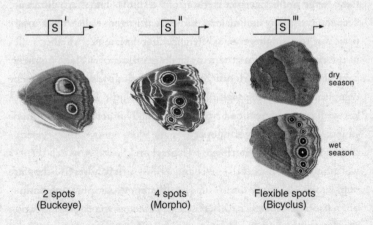

2 spots
(Buckeye)

4 spots
(Morpho)

Flexible spots
(Bicyclus)

FIG. 8.8 **Modification of the *Distal-less* eyespot switch explains different eyespot patterns.** The evolution of eyespot number (S′ and S″) and the control of eyespot size (S‴) has occurred by changing switches in different ways. DRAWING BY LEANNE OLDS

Mimicry and Color-Pattern Evolution

The changing color patterns of other wing-pattern elements dominate the evolutionary story in most butterflies. The difference in appearance between species, or between individuals of the same species, is due to different spatial patterns of pigment synthesis and scale structural colors. While every species is a story, I will close this chapter by returning to the evolution of mimicry that has played a large role in the discussion of natural selection, and which remains, from the viewpoint of Evo Devo, a mystery yet to be solved.

Striking differences in appearances can often have relatively simple genetic and developmental bases. For example, in the tiger swallowtail (*Papilio glaucus*) of eastern North America (including Wisconsin!), two forms of the female occur: a yellow form with black tiger stripes, and a black or "melanic" form (plate 8j). This latter type is a specific mimic of the pipevine swallowtail (*Battus philenor*), which flies in the same range as the tiger but is distasteful to birds. The sharp difference between the yellow and melanic forms of the tiger swallowtail appears to be due to a single genetic difference that determines whether scales in the central region of the wing make a yellow or a black pigment. Even though the pattern differences between individuals appear to be complicated, involving multiple pattern motifs, the genetic differences between forms appear to be relatively few. This also seems to be the case in other examples of mimicry.

The *Heliconius* butterflies of Central and South America display warning colors, especially reds and yellows, that advertise they are unpalatable. Mimicry occurs among different geographic populations of *Heliconius* butterflies. Different species in a given geographic region converge on a similar wing pattern, but different geographic populations of these same species may display different patterns. For example, *H. melpomene* and *H. erato* appear similar to each other in each locale in Brazil, Ecuador, and Peru, but the different geographic varieties of each species bear different sets of markings (plate 8k). The general idea is that the birds that attack these butterflies differ from region to region and each butterfly species adapts under this selection pressure toward a form that is most effective at discouraging attack by local predators. Extensive genetic studies have been undertaken to attempt to ascertain the genetic differences contributing to the size, shape, and color of bands and rays on *Heliconius* wings. In general, a modest number of genetic differences appear to control the differences between populations.

Biologists have not yet precisely identified the genes involved in swallowtail or *Heliconius* color patterns and mimicry. But this is now only a matter of time. When these genes are identified, a great opportunity will arise to make the connections among fitness, genes, and the forms of these marvelous patterns.

While revelations about the mysteries of the evolution of the Technicolor patterns of butterfly wings lie ahead, biologists have recently made some great discoveries about the evolution of one color—basic black—in a host of other animals. The evolution of black or melanic forms is one of the most prevalent color changes in the animal kingdom. In the next chapter, I will tell how the study of black coloration has enabled biologists to capture evolution "in action."

An African scene. JAMIE CARROLL

9

Paint It
Black

> Fortunate are those who have learned to see, in the
> wild things of nature, something to be loved, some-
> thing to be wondered at, something to be rever-
> enced, for they will have found the key to a
> never-failing source of recreation and refreshment.

> —Hugh B. Cott, *Adaptive Colouration in
> Animals* (1940)

"WHAT IN THE WORLD have you been doing to yourself, Zebra?
Don't you know that if you were on the High Veldt I could see you ten
miles off? You haven't any form."

"Yes," said the Zebra, "but this isn't the High Veldt. Can't you see?"

"I can now," said the Leopard. "But I couldn't all yesterday. How is
it done?"

"Let us up," said the Zebra, "and we will show you."

They let the Zebra and Giraffe get up; and Zebra moved away to some little thorn-bushes where the sunlight fell all stripy, and the Giraffe moved off to some tallish trees where the shadows fell all blotchy.

"Now watch," said the Zebra and Giraffe. "This is the way it's done. One—two—three! And where's your breakfast?"

Leopard stared, and Ethiopian stared, but all they could see were stripy shadows and blotched shadows in the forest, but never a sign of Zebra and Giraffe. They had just walked off and hidden themselves in the shadowy forest.

"Hi! Hi!" said the Ethiopian. "That's a trick worth learning. Take a lesson by it, Leopard."

Although much of the reading world was charmed by Rudyard Kipling's story of how zebras concealed themselves in "How the Leopard Got His Spots," apparently Teddy Roosevelt was not. Shortly after the end of his second term as president in 1909, he set out on a yearlong hunting safari in Africa. In his account of that journey, *African Game Trails* (1910), Roosevelt assailed contemporary views of animal coloration:

> Very much of what is commonly said about "protective coloration" has no basis whatever in fact. . . . Giraffes, leopards, and zebras, for instance, have actually been held up as instances of creatures that are "protectingly" colored and are benefited thereby. The giraffe is one of the most conspicuous objects in nature . . . it is safe to say that under no conditions is its coloring of the slightest value to it as affording it "protection" from foes trusting to their eyesight. So it is

with the leopard; it is undoubtedly much less conspicuous than if it were black—and yet the black leopards, the melanistic individuals thrive as well as their spotted brothers . . . the leopard's coloration represents a very slight disadvantage, and not an advantage to the beast; but its life is led under conditions which make either the advantage or the disadvantage so slight as to be negligible . . . it is normally nocturnal and by night the color of its hide is of no consequence whatever.

He reserved his greatest cynicism for zebras:

All of this applies with peculiar force to the zebra, which it has also been somewhat the fashion of recent years to hold up as an example of "protective" coloration. As a matter of fact the zebra's coloration is not protective at all. On the contrary, it is exceedingly conspicuous and under the actual conditions of the zebra's life probably never hides it from its foes; the instances to the contrary being due to conditions so exceptional that they may be disregarded.

The great hunter added: "The truth is that no game of the plains is helped in any way by its coloration in evading its foes and none seeks to escape the vision of its foes. . . . On the plains one sees the wildebeest farthest off and with most ease; the zebra and hartebeest next; the gazelles last." And he challenged: "If any man seriously regards the zebra's coloration as 'protective,' let him try the experiment of wearing a hunting suit of the zebra pattern; he will speedily be undeceived."

I don't think I would have recommended putting on a zebra suit anywhere near Roosevelt's hunting party. Of the 512 animals he and his son Kermit shot that year, 29 were zebras, more than almost any other species.

Zoologist Hugh B. Cott, who spent much more time in Africa than Roosevelt, did not agree with the President. He based his now classic encyclopedic volume *Adaptive Colouration in Animals* (1940) on extensive research. Cott was also a talented artist whose mastery of drawing techniques provided him with even greater insight into how color patterns conceal, advertise, or disguise animals. Cott's expertise was by no means just an academic indulgence. His great book was written on the eve of Britain's engagement in World War II, during which Cott served as a "camoufleur," advising the military on the design of better camouflage.

Zebras, Cott explained, use the principle of disruptive patterns to obliterate their contours (figure 9.1):

> In the dusk, when he is liable to be attacked, and in country afford-ing thin cover, he is one of the least easily recognized game animals. White, who claims a wide experience with these animals, and states that he has seen "thousands upon thousands" of zebra against dif-ferent backgrounds, writes that: "At any rate, in the thin cover described he is the most invisible of animals. The stripes of white and black so confuse him with the cover that he is absolutely unseen at the most absurd ranges."

FIG. 9.1 **Disruptive coloration in zebras.** Vertical stripes help conceal the contours of the animal in backgrounds of alternating light and shadow. FROM HUGH COTT, *ADAPTIVE COLOURATION IN ANI-MALS* (LONDON: METHUEN AND CO., 1940); USED BY PERMISSION

In addition to this concealment theory, there are other explanations for zebra stripes. In the herd, an individual is difficult to make out against the backdrop of all those other moving stripes. Perhaps this conveys some advantage in confusing predators. Another theory is that the striped pattern may reduce the biting of insect pests, which some believe prefer animals that are dark all over. Yet another possibility is that these patterns aid mother and young zebras in identification, or individuals in finding the herd (zebras are apparently drawn toward black and white stripes painted on boards).

These ideas and Cott v. Roosevelt illustrate that, in biology, answers to questions about the purpose of patterns are often not black and white. Anecdotes inspire hypotheses, but they do not furnish reliable conclusions. In Kenya, I twice came right up to trees without spotting the leopards reclining in them. I also watched a leopard descend and stalk a dik-dik while concealed in the brush in broad daylight—contrary to Roosevelt's assertion about leopard's nocturnal habits. The advantages or disadvantages of having or not having spots, of being striped or solid, black or white, can be decided only by controlled data, which, as you can imagine, might be very difficult to obtain in many circumstances.

Nevertheless, it is obvious that the coloration of animals plays a critical part in how they interact with other species and their own kind. The natural history of coloration has therefore held an important place in evolutionary biology, particularly as examples of natural selection and sexual selection. What has always been missing in these examples is knowledge of the precise genetic and developmental basis of traits. But now, thanks to molecular biology and Evo Devo, there are a handful or more of examples where the mechanisms underlying the differences within or between species are well understood—in some cases, right down to the precise DNA changes responsible, the "smoking guns" of evolution.

In this chapter, I will focus on the evolution of just one color: black. The evolution of dark coloration of body parts or of the whole body constitute some of the most frequent changes in Nature. I will highlight how black coloration has evolved in jaguars, birds, pocket mice, fruit flies, and a handful of domesticated species. In some cases, we know both the selective forces involved and the molecular origin of the evolution of a trait. This connection between form, fitness, and specific genes bridges a critical gap in the Modern Synthesis. Thus, the stories here are important new "icons" of evolution and should join the classic cases of peppered moths and Galapagos finches as examples of the evolutionary process.

I will also discuss how the new vantage point provided by Evo Devo allows us to address questions that were nearly impossible to answer before. A few of the most intriguing questions are: Does evolution ever repeat itself? Does the same genetic change ever occur independently in different species? Or, is there more than one path to similar evolutionary adaptations?

Melanism in Nature

The term *melanism* refers to the condition whereby an individual or species displays broader areas or greater amounts of black or dark coloration in place of other colors. Melanin pigments are complex chemical polymers that occur in many forms with different hues, from basic black to brown, reddish brown, buff, or tan.

The occurrence of melanism is widespread throughout the animal kingdom and has been well studied in groups of insects (especially moths and ladybird beetles), land snails, mammals, and birds. Melanic pigmentation can serve many roles, including protection from ultraviolet light damage, thermal regulation (it is often found at high altitudes and helps warm creatures more quickly), camouflage and concealment, in mate choice, and other functions. Because of this wide array of possibilities, it is often difficult to say precisely and exclusively what a given melanic pattern is "for."

Michael Majerus has noted that melanic forms are known of perhaps half the moth species in the United Kingdom. The most famous example of melanism and natural selection is the evolving distribution of the peppered moth (*Biston betularia*) in industrial areas of England and the northern United States over the past 150 years or so. The moth has two distinct forms: the *typica* form, which is white with black speckling, and the *carbonaria* form, which is almost completely black; there is also a range of intermediate forms (figure 9.2).

In industrial areas, pollutants kill lichens on trees and soot blackens the trunks. The consequence is that in industrial areas, the *typica* form is conspicuous and the *carbonaria* form is camouflaged on these backgrounds. A combination of observations—on the frequency of the two forms, field study of their resting site preferences, and bird predation—has led to the general conclusion that the changes in frequency of the two forms in time and across regions are due to differences in the natural selective pressure of bird predation. While there have been some critiques of some of the methodology used in individual studies of industrial melanism in the past, the peppered moth story stands as a classic case of evolution observed in real time.

The genetics of melanism in the *carbonaria* form appears to be relatively simple. Crosses of the two types suggest that one major gene determines the bulk of the difference in pigmentation and a few other genes modify the degree of melanism. The exact identity of the gene involved in industrial melanism, however, has not yet been revealed. This would be a nice capstone for a 150-year-old story. Ironically, stricter pollution laws, which are surely a positive development, have doomed the *carbonaria* form and it may disappear altogether within the next couple of decades, so the molecular geneticists had better get on the ball.

In several other species, the genes involved in melanism have been identified and I will concentrate on those examples for the remainder of this chapter.

FIG. 9.2 **Melanism in peppered moths.** The light form is obvious on a black background (top) and the melanic form is cryptic on a dark background (middle). The mottled light pattern is cryptic on a lichen-covered tree trunk (bottom). PHOTOS COURTESY OF TONY LIEBERT AND PAUL BRAKEFIELD, UNIVERSITY OF LEIDEN

How the Jaguar Covers His Spots

Melanic forms of big cats are well-known and chances are that you have encountered a black leopard in a zoo somewhere. Black leopards are very rare on the African savannah, but in the jungles of Southeast Asia the melanic form is often the most common. The dark coat color may be an advantage in escaping detection by potential prey. Black jaguars are also known from reports throughout most of their range in Central and South America. While these cats are referred to as "black," their spotting is still evident (figure 9.3). The dark coloration appears then to be superimposed upon the orange and black pattern, but does not obliterate it.

In mammals, two types of melanin are produced by the pigment cells of the skin and the hair follicles, eumelanin and phaeomelanin,

FIG. 9.3 The orange and melanic phases of the jaguar. COPYRIGHT NANCY VANDERNEY, EFBC/FCC; USED BY PERMISSION

which are responsible for the black-brown and reddish orange (or yellow) coloration of fur, respectively. The amounts of each pigment made are controlled by several proteins. One pivotal protein is called the melanocortin-1 receptor, or MC1R. This protein sits on the cell membrane of pigment cells, with one part extending outside the cell, and another extending inside the cell. A hormone called alpha-melanocyte-stimulating hormone (MSH) binds to the MC1R protein; this triggers a cascade of events inside the pigment cell that leads to the production of eumelanin-synthesizing enzymes. There is also a protein called Agouti that blocks the receptor, and when that occurs, phaeomelanin is made. Pigment type therefore depends upon the state of activity of the MC1R protein (figure 9.4).

Examination of the gene-encoding *MC1R* of normal orange-phase jaguars and melanistic jaguars has revealed that a specific mutation exists in this gene exclusively in all black jaguars. This mutation deletes five amino acids and changes another amino acid in the MC1R protein. A cat that has one copy of the *MC1R* mutation and one copy of the normal *MC1R* gene is black; melanism is therefore said to be dominant. This means that the mutant form of the protein overrides the presence of the normal form. The change in the MC1R protein has caused it to be continuously active in stimulating eumelanin synthesis; it is blind to the presence of either hormones or inhibitors.

This is the first time in this book that I have described a change in a protein's sequence (as opposed to that of a genetic switch) that is clearly responsible for a difference in appearance between animals. The reason the MC1R can change is that this receptor is largely dedicated to the regulation of pigment synthesis. Changes in its activity do not compromise other body functions. The melanocortin-1 receptor is a member of a family of five receptors that have specialized functions in many facets of mammalian physiology, and that respond to a related family of hormones. The ability of pigmentation to evolve without affecting other functions is due to the evolution of

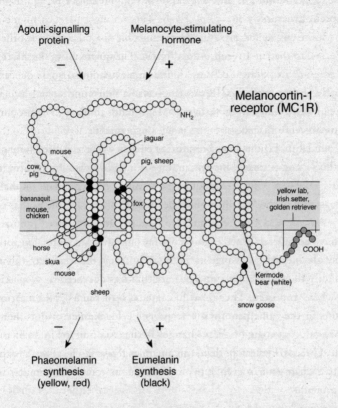

FIG. 9.4 **Changes in the melanocortin-1 receptor protein are associated with melanism in mammals and birds.** The MC1R receptor spans the membrane of melanocytes. It is stimulated to produce black eumelanin by the alpha-melanocyte-stimulating hormone and inhibited by the Agouti protein, which switches to promote phaeomelanin synthesis. The dark circles indicate positions in the protein that are associated with melanism in various species. Changes at the tail of the molecule are associated with white coloration in Kermode bears, and different varieties of dog coat colors. DRAWING BY LEANNE OLDS; ADAPTED FROM M. MAJERUS AND N. MUNDY, *TRENDS IN GENETICS* 19 (2003): 585; REPRINTED WITH PERMISSION OF ELSEVIER

MC1R regulation, in that it is expressed in pigment cells, and of its protein structure.

Mutations in *MC1R* are also responsible for melanism in other species. In the jaguarundi, a smaller cat that ranges from the southwestern United States to South America, melanism is also associated with a change in the MC1R protein, but in a slightly different place in the protein. Thus, in these two cats, dramatic changes in coat color have arisen from changes in the same, single protein.

MC1R mutations have been found to also cause melanic plumage in birds. For example, the bananaquit is a widely distributed bird in the Caribbean. Most bananaquits have bright yellow plumage on their breast and a stripe of white plumage along their eye. On the islands of St. Vincent and Grenada, melanic bananaquits are found that are almost completely black (figure 9.5). This melanism is associated with a single amino acid change in the MC1R protein. Interestingly, this is exactly the same amino acid change that has occurred in domestic chickens and mice. Thus, in wild cats and birds, independent mutations in the same protein are responsible for similar evolutionary changes, and some of these changes have also occurred in domesticated breeds. Evolution, then, can and does repeat itself, at the level of a particular gene or even at the level of the same single amino acid in a protein.

FIG. 9.5 Melanism in bananaquits. PHOTOS COURTESY OF ANDREW MACCOLL

While we know that the melanism in domestic species was selected for by humans, we do not know the selective pressures that affect the frequency of melanism in wild cats or bananaquits. For the predators, black coloration could help avoid detection while hunting, and in the bananaquit it may affect habitat choice at different elevations, but these are just speculations. In another species, however, the selective advantage or disadvantage of melanism is clear and these animals present a surprising twist on the MC1R story in the evolution of melanism.

Rock Pocket Mice: More than One Way to Paint Them Black

The deserts of the southwestern United States include many varieties of local habitats that impose different demands on the animals and plants living within them. As such, this vast area provides a great laboratory for understanding evolutionary adaptations.

In the Pinacate region of southwestern Arizona, there are dark rocky habitats that are the products of lava flows from less than a million years ago. The rock pocket mouse, *Chaetodipus intermedius*, inhabits this region as well as other rocky areas of the southwest. Naturalists in the 1930s noted that mice found on these lava rocks are typically melanic while those found in surrounding areas of light sandy soil are usually light-colored (figure 9.6). The correlation between habitat preference and fur color is thought to be an adaptation against predators, particularly owls. These birds are well documented to feed on these mice, and experiments have shown that owls can discriminate between dark and light mice, even at night (the clear skies of the desert allow substantial moonlighting of the rocks and soil). Further support for this adaptive value of fur color comes from the observation of similar patterns of the distribution of mice at numerous localities.

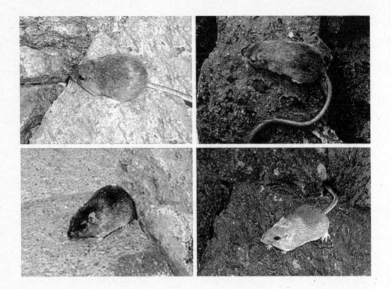

FIG. 9.6 **Association of habitat and fur color in rock pocket mice.** Light-colored mice are typically found on light-colored rocks and dark mice on dark lava flows, which affords them protection from predators. PHOTOS COURTESY OF MICHAEL NACHMAN, FROM M. NACHMAN ET AL., *PROCEEDINGS OF THE NATIONAL ACADEMY OF SCIENCE, USA* 100 (2003): 5268; USED BY PERMISSION

In order to get at the genetic basis of melanism in these mice, Michael Nachman and his colleagues at the University of Arizona recently examined the sequence of the *MC1R* genes of both light and dark mice. They found a perfect correlation between the presence of four mutations in the *MC1R* gene in dark mice that resulted in four amino acid differences from the light mice, and the MC1R protein of other mice. These differences indicate that in this population, as in the jaguar, jaguarundi, and bananaquits, a continuously active form of MC1R exists that produces black coat color. The genetic evidence, the distribution of the differently colored mice, and additional field and molecular evidence all provide a compelling picture of how the appearance of an animal can evolve under selection in a natural setting.

This is a very satisfying story, but it doesn't end there. Nachman and his colleagues also examined light and dark rock pocket mice from a second locality, about 475 miles away from the Arizona population, living on and around lava flows in New Mexico. While the ecological story is the same, the genetics are different. The dark mice in New Mexico did not carry any *MC1R* mutations. Nor did they carry mutations in another gene, *Agouti*, that will also cause black coloration when mutated (because *Agouti* inhibits MC1R, mutation in the inhibitor allows MC1R to be fully active). This means that there are other genes besides *MC1R* and *Agouti* that can be mutated to cause melanism. Thus, the two different melanic populations of the same species, living in similar terrain, on rock flows less than a million years old, have found different means of evolving melanism. Evolution need not always follow the exact same genetic path, even in the same species.

Black Leopards, White Bears, and Redheads

It is also the case in leopards and some other wild cats that melanism has evolved through mutations in other genes besides *MC1R* or *Agouti*. We know from many species that fur color is affected by more than just these two genes, and by specific combinations of pigmentation genes. For example, the coat colors of yellow Labradors, golden retrievers, and Irish setters appear to be due to a mutation that eliminates *MC1R* function, with other genes determining the yellow, orange, or red coat color. Biologists are working to identify these other genes involved in melanism and coat color, and they are

FIG. 9.7 A Kermode bear and a black bear. PHOTO COURTESY OF
CHARLIE RUSSELL

sure to find the specific changes responsible for differences between
species and breeds.

While melanism is often due to mutations that activate *MC1R*, other
types of mutations in *MC1R* are responsible for some other notable
color patterns. The white "Kermode" or "spirit" bear of the Pacific
Northwest was once thought to be a separate species, but it is actually
just a color variant of the black bear (figure 9.7). The Kermode bear
carries a mutation in the *MC1R* gene that knocks out the function of
the receptor. Because *MC1R* is crippled, no black pigment is made and
the bear has a white coat.

Finally, in humans, mutations in *MC1R* are responsible for red hair.
This mutation is also responsible for the freckling, light pigmentation,
and sun sensitivity of these individuals. Tanning in humans is due to
the production of eumelanin in response to ultraviolet light stimulation and is regulated by the stimulation of MC1R by alpha-MSH.
Mutations in *MC1R* that cause red hair color in humans appear to
reduce the ability of MC1R to respond to alpha-MSH.

The Evolution of Fancier Patterns in Mammals: Stripes and Spots

The examples of fur color and feather plumage I have described generally involve the entire coat or feather pattern. We have seen that the evolution of all black, white, red, or yellow coats involves mutations in pigmentation genes, most often *MC1R*. But wild fur and plumage patterns are more often made up of two or more colors in some spatial pattern. This means that the expression of pigmentation genes must differ in areas of the body that will be of different colors. In order to selectively express those genes in one place and not another, there must be switches that control the expression of pigmentation genes and paint-color patterns.

In mammals, our understanding of these switches is just beginning. One of the most common coat-color patterns in mammals is a buff, brown, or dark coloration of the back and sides, with light coloration of the underside. This is, in fact, the color scheme of the wild house mouse. The *Agouti* gene is central to making the underside and topsides of the animal different in appearance. There is a specific genetic switch that drives *Agouti* expression in the hair follicles on the underside of the animal. Because the Agouti protein inhibits MC1R activity, that makes this fur lighter in color.

While we know some details about how solid and two-toned fur patterns in mammals are formed, how about fancier color patterns, such as those stripes on a zebra? In one of my all-time favorite essays by the late Stephen Jay Gould, he addressed the question "Is a zebra a white animal with black stripes or a black animal with white stripes?" This riddle of natural history has inspired many comments over the years. The balance of opinion now leans toward the black animal/white stripes verdict. But before I lead you down any path, let me first just focus on the question of "How does a zebra get its stripes?"

The truth is that, compared with everything else I have told you about in this book, we don't really know the answer directly. To my knowledge, no one has studied zebra embryology. But we can try to piece together a scenario by combining snippets of different types of information available for other mammals. This includes knowledge of how the cells that make melanin develop in embryos, observations on coat-color mutants in mice, horses, and other mammals, the appearance of hybrids resulting from the breeding of zebras with horses, and the variation in stripe patterns within and between zebra species.

The most important clue to the origin of striping is the origin of the pigment cells called melanocytes that will color those stripes. These cells arise from precursors in what is called the neural crest, a region near the spinal cord. The precursors, called melanoblasts, stream out of the neural crest and migrate along tracks that are generally perpendicular to the spinal cord. The migrating cells appear to be following some guidance cues on their journeys. Since the tracks start at the very top of the back and run down, the belly and chest regions are the last places they reach. It is very clear that mutations that slow or reduce melanocyte migration will leave these areas white. This is the basis for white paint on horses, white chests on dogs, and white bellies on cats.

In zebras then, the strong black stripes on the coat are regions into which melanocytes must have migrated. What is not known is whether the white stripes are regions that lack melanocytes (either because they didn't migrate into these areas, or died off), or whether these areas have melanocytes that are inhibited from producing pigment. Whether the difference between black and white areas is due to migration, death, or inhibition of melanin production, each of these mechanisms would require the regulation of a specific process in a striped pattern. For example, in the first case we know of many signaling molecules that are expressed in striped patterns in the neural tube and somites of vertebrates. If migrating melanoblasts were channeled or retarded by any of these molecules, then a striped pattern of melanoblasts would result. Alternatively, if an inhibitor of melanin production was expressed in a striped pattern in the skin or follicles, this could also lead to striped fur patterns. Because the bellies of zebras are generally white, I favor the explanation that the pure white color of the stripes is due to the absence of melanocytes. However, even if this guess is correct, there are multiple ways for the melanocytes to be missing; the exact developmental mechanism remains a wide-open question.

So is a zebra black on white or white on black? Here's just one more interesting tidbit to help us decide: rare zebras have been described that are white-spotted where stripes would normally appear. This is what one would predict if the "default" coat pattern is all black. But I don't think this "default" idea is necessarily the way to look at a zebra. Furthermore, in March 2004, the Kenya Wildlife Service reported the birth of an all-white zebra foal. From a developmental point of view, both the black stripes and the white stripes are actively being "drawn." I'd prefer to say that a zebra is an animal with both black and white stripes.

The migration of melanoblasts out of the neural crest perpendicular to the spinal cord suggests, in fact, that the potential to be striped is an inherent property of this developmental process. Mutations in animals that are usually solid-colored, such as mice and horses, can lead to striping. Furthermore, horse and donkey breeding lore is full of examples of partially striped animals, such as the brindle pattern. Darwin himself took extensive note of striped horses and donkeys in *The Origin of Species*, particularly in hybrids. It is possible to make horse-zebra hybrids by mating zebras sires with horse mares. The offspring are typically striped; however, if the mare had white paint, the black stripes appear only on the dark regions of the hybrid animal. This is consistent with the white paint gene affecting melanoblast migration such that stripes occur only where melanoblasts have migrated. Perhaps even more provocative is the observation that horse-zebra hybrids usually display *more* stripes than the zebra parent.

Jonathan Bard has taken note of this curiosity as well as the difference in stripe number among the three living species of zebras to come up with a very intriguing model for all zebra patterns. Bard noted that Grevy's zebra has about eighty stripes, mountain zebras about forty-three, and the common zebra twenty-five to thirty stripes. He proposed that these different numbers of stripes are determined by differences in the time in development at which melanocytes begin their journeys in each species. Bard noticed that stripes were wider in zebras with fewer stripes, narrower in zebras with more stripes. He proposed that this relationship could be explained if stripes were initiated at constant intervals by some mechanisms in early zebra embryos (about every 0.4 millimeters) but at different times in different species (figure 9.8). The earlier the stripes are initiated, the larger the stripes would be and the fewer stripes overall would fit onto the animal. Conversely, the later the stripes were initiated, the smaller each stripe would be relative to the whole embryo, but more stripes would fit onto the animal. Bard reasoned that the cause of the larger number of stripes in zebra-horse hybrids was that stripe formation was delayed in the hybrid relative to the zebra parent (a very reasonable idea because hybrids often develop a bit more slowly than their parents).

A crucial element of Bard's model is that the striping process of the zebra is initiated when the embryo is very small, six months before pigmentation of the fur actually begins. This is a very important point about the making of patterns in what will be large animals. The developmental processes for making patterns operate over modest distances in terms of how far away cells can be and still communicate with one another. The distances between stripes in newborn and older zebras are too great for cells in these stripes to be communicating with those in another stripe. The general outline of animal patterns is defined early; they then grow larger with these as yet invisible designs in place.

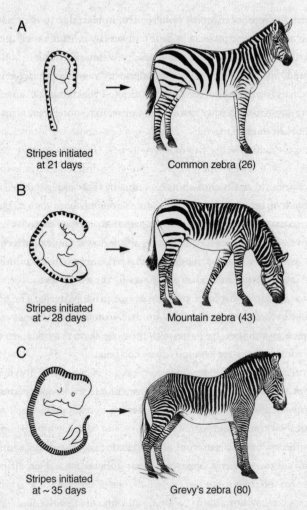

A

Stripes initiated
at 21 days

Common zebra (26)

B

Stripes initiated
at ~ 28 days

Mountain zebra (43)

C

Stripes initiated
at ~ 35 days

Grevy's zebra (80)

FIG. 9.8 **Jonathan Bard's model for the generation of different
numbers of stripes in different zebra species.** Bard suggested that
if stripes are generated at the same interval (every twenty cells) at
slightly different times in different species (A, B, and C), it will produce the different numbers and width of stripes in the common
zebra, mountain zebra, and Grevy's zebra. DRAWING BY LEANNE OLDS;
MODIFIED FROM J. B. BARD, *JOURNAL OF ZOOLOGY* 183 (1977): 527

If the differences in stripe number are, in fact, due to the relative timing of the stripe process in different species, this must be due to shifts in the timing at which the genes for melanoblast migration are activated. Shifts in timing are fundamentally regulatory changes, so differences in stripe number must be due to evolutionary changes in genetic switches that control the timing or spatial patterns of melanoblast migration.

What about the making of spots? As much as I would like to tell you how the leopard gets his pattern, there is even less hard data to go on for these patterns, in mammals at least, than for striped patterns.

More is known, however, in insects about how complex patterns of black spots and stripes are made, and this has been a particular interest in my laboratory. The bodies and parts of the many species of fruit flies, for example, display a great variety of black patterns. The black pigment in these bugs is also melanin. In *Drosophila melanogaster* the abdomen and thorax are patterned, the bristles on the body are very dark, but the wings are generally clear and pale.

In other species, large amounts of black pigment may be distributed throughout the body, or restricted to specific places. In one species, *D. biarmipes*, the wings of male flies bear a conspicuous black spot toward their tips (figure 9.9). This spot is used in courtship displays in which the male prances in front of the female and extends his wings so she can see the spot and, apparently, that does the trick. Hey, different strokes for different folks . . .

In species without spots, a small amount of a particular black-pigmentation-producing protein is made in all wing cells. But in *D. biarmipes*, much greater amounts of this protein are made in cells that will form the wing spot. We believe that this difference may be due to evolutionary changes in a switch that controls how this protein is expressed in fruit fly wing cells. Pigmentation genes have switches that

FIG. 9.9 **Wing spots on fruit flies.** These spots play a role in courtship in species that bear them. Expression of pigmentation genes makes the differences between species. PHOTOS BY NICOLAS GOMPEL

control their expression in different body parts. The independence of these switches allows one part of the body to evolve a new pattern, independent of patterns on other body parts. Based on what we've seen in flies, I predict that birds, mammals, fish, snakes, and other animals have evolved switches that also control their coloration genes, and that much of the diversity of body-color patterns in these groups is due to evolutionary changes in these switches.

Selection, Genes, and Fitness: How Much of an Advantage Matters?

In the past two chapters I have mentioned arguments and evidence for natural selection on butterfly eyespot patterns, melanistic cats, dark and light rock pocket mice, and zebra patterns, as well as sexual selection for spots on fruit fly wings. It may be obvious in some cases how these color patterns provide an advantage to those individuals that bear them. But how much of an advantage is sufficient in order that a selection will favor these individuals? In 1910, Teddy Roosevelt couldn't see how the leopard's spots or zebra's stripes provided an advantage. I think that, like many people, Roosevelt believed that the advantage must be easily seen or measured for selection to favor one pattern over another. The key question then is: How much of a difference must there be to matter?

This is the realm of population genetics, the branch of genetics that is concerned with variation among individuals, the genetic basis for it, and the changes in the frequency of forms and genes in evolution. The short answer to the question "How much of a difference matters?" is that natural selection requires a surprisingly small difference in relative success between two forms to work. This difference may often be imperceptible or unmeasurable in the field, but is sufficient to favor the evolution of one form over another.

Population geneticists have developed some formulas that reveal the basic relationships between the advantage or disadvantage of a particular mutation and the fate of that mutation in a population or species. Using these formulas we can ask how much better one form has to be to take over a population. And, how long does that take?

There are several factors and concepts to consider. When we say "better," what do we mean? The concept is called "fitness," and for animals that is really a composite of survival (how long an individual is around) and fecundity (how many offspring it produces). For selection to work and a new mutation to prevail, it must provide some relative advantage in fitness. Suppose, for example, that individuals with a new mutation (say, melanism in a moth or in a rock pocket mouse) leave on average 101 offspring versus 100 for individuals without the mutation. That is a difference in relative fitness of just 1 percent. In our formula we will convert this figure to a *selection coefficient* called s of 0.01.

Can that slight a difference matter? If that 1 percent advantage is maintained, you bet it does. A mutation will increase in frequency in a population at a rate that depends upon the size of the population and the magnitude of the selection coefficient. The formula for determining the time, in generations, for a mutation to spread throughout a population is:

Time = 2/s ln(2N), where N = the number of individuals in the population and ln is the natural logarithm

In our example, if N = 10,000, which is a large and reasonable number, and s = 0.01, then it will require 2/0.01 ln (2 × 10,000) = 1980 generations. In a mouse or a moth, this would represent 2000 years, or less. If s = 0.001, meaning that there is just a 0.1 percent advantage, that mutation will still become fixed in about 20,000 generations. These calculations show that even with a very small advantage, a mutation will spread in a geologically brief period of time. Selection coefficients need not be so small, however. For melanic moths in industrial England, or insects resistant to insecticides, the rapid rise in the frequencies of individuals was measurable over periods of years, not millennia. The selection coefficients in these cases have been estimated to range from 0.2 to 0.5, which are quite large and reflect huge selective advantages.

In contemplating the power of selection, we must also consider the converse to the spread of adaptive mutations—that is, the elimination of mutations that are disadvantageous. I won't go into the mathematics here, but suffice it to say that mutations that cause even a slight disadvantage have very, very little chance of spreading throughout a large population. When considering the advantages or disadvantages of black or white fur on a rock pocket mouse, we should also think about what patterns we don't see in nature, such as spotted rock mice. Should such a mutation arise, these animals would stand out against either light or dark backgrounds. The absence of this form could be because that mutation never arises, but I don't think that's the reason. It is more likely that when such individuals do occur, they are at such a disadvantage that they never thrive in any significant numbers in the wild.

Let's close this chapter by returning to the zebra debate using the same logic I just applied to spotted rock pocket mice. In deciding the value of stripes, isn't it worth thinking about why all those zebras we see are striped? If it didn't matter, wouldn't we see lots of zebras without stripes? Indeed, we should. Fur coloration mutants are sufficiently common in mammals that dramatic mutants (e.g., white tigers or spotted zebras) occur at some rare frequency in the wild. And in domesticated animals, breeders over the ages have selected rare varieties that arise spontaneously, such as the many coat-color patterns of horses, a relative of zebras. I submit that the laboratory of the African plain is telling us that stripes do matter.

We just do not know what purpose they serve. Take your pick of theories—the important fact to keep in mind is that just the slightest relative advantage to being unstriped is all that is needed for the stripes to remain. The fundamental principles of the power of natural selection on gaining or maintaining a character (including sexual selection) pertain to the evolution of all species, including our own. So, too, do the fundamental lessons of Evo Devo about modularity, genetic switches, and the evolution of form. I will turn, at last, to the making of *Homo sapiens* and its characteristic features in the next chapter.

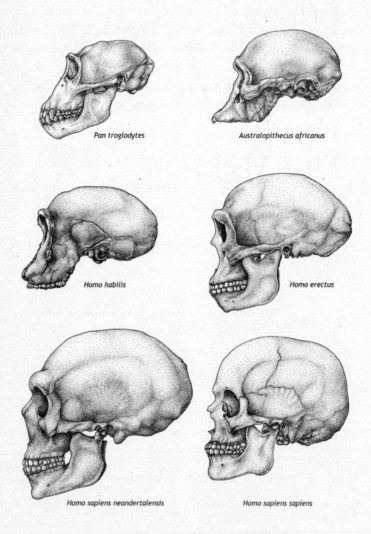

Pan troglodytes

Australopithecus africanus

Homo habilis

Homo erectus

Homo sapiens neandertalensis

Homo sapiens sapiens

Evolution of hominid skull size and patterns. DRAWING
COPYRIGHT BY DEBORAH J. MAIZELS, ZOOBOTANICA

10

A Beautiful Mind: The Making of *Homo sapiens*

The difference in mind between man and the higher animals, great as it is, is one of degree and not of kind.

—Charles Darwin,
The Descent of Man (1871)

SOON AFTER HIS RETURN from his around-the-world voyage, Darwin visited Jenny, the first orangutan displayed in the London Zoo and among the first apes ever displayed in Britain. She made a profound impression on the naturalist. He was astonished at how she interacted with her keeper, and he admired her playfulness and intelligence.

Her emotions appeared to be those of a child, and from that first encounter onward, Darwin would look at children, including his own, as a comparative primatologist.

Close encounters with apes can be as unsettling as they are fascinating. Queen Victoria, after viewing another orangutan (also named Jenny) wrote that she was "frightful, and painfully and disagreeably human."

In the expressions on the faces of chimps, orangutans, and gorillas, their mannerisms, and their beautiful, dexterous hands, we see reflections of ourselves. These reflections have always raised provocative and, for some, discomforting questions about the gap between man and beast. What do the apes see when they glance toward their hairless, bipedal visitors? What is going on behind the long stare of a gorilla? What rolls of the ecological and genetic dice put us on the outside of those enclosures looking in, and not the other way around?

My fourteen-year-old niece, Caitie, impressed by an ape exhibit in Tampa, Florida, turned to her father and asked, "You are always telling us that we are 99 percent identical to chimpanzees. Okay, but what makes us *different*?"

Excellent question.

Caitie was referring to the often quoted figure of our nearly 99 percent identity at the DNA sequence level to chimpanzees. In this chapter, I will frame the beginning of an answer to her question. I have to say "beginning" for two reasons. The first is that biology has just arrived at the point where we can explore the question of specific genetic differences between ourselves and apes. Many more discoveries lie ahead than are yet in hand. The second is that certain kinds of data, such as the visualization of gene expression patterns, which has taught us so much about the evolution of animal form, will be scarce for human embryos.

The psychologist Erich Fromm once said, "Man is the only animal for whom his own existence is a problem he has to solve." It is clear that this solution requires an integrated picture that encompasses many areas of science, including traditional fields such as paleontol-

ogy and comparative neuroanatomy that have long sought to understand human history and the biological basis for our mental faculties,
as well as emerging disciplines, such as comparative genomics, human
medical genetics, and Evo Devo, that are just now taking the stage.

The changes in human form and function that have occurred in the
6 million years since the last common ancestor we shared with chimps
are the product of the evolution of human development and genes.
Understanding how the features of greatest interest evolved—such
as our skeleton (bipedalism, limb lengths, hand and thumb, pelvis,
and skull), life history (gestation time, prolonged juvenile state, and
longevity), and, most especially, our larger brain, speech, and language—
presents some of the greatest puzzles in biology, and for Evo Devo in
particular.

In this chapter, I will examine the evolution of human form from several perspectives—the fossil record, comparative neurobiology, embryology, and genetics—and explore four big questions of these fields:

1. What was the actual pattern of human evolution in terms of
   the changes that occurred among species leading to modern
   humans?
2. Was human evolution in any way atypical of other mammals?
3. Where in our brain do human capabilities reside?
4. Where in our DNA are the differences that distinguish us from
   other apes?

The central message of this chapter is that what we have learned
thus far about the evolution of form in other animals—butterflies and
zebras, fruit flies and finches, spiders and snakes—fully applies to the
evolution of human form. Our physical evolution was no different
than that of other species. The evolution of human features—
including our upright posture, large brain, opposable thumb, speech,
and language—is due to developmental changes that modified existing
primate or great ape structures, and that were accumulated over several

million years and many speciation events. Some of the specific genetic differences between ourselves and living apes are now being revealed.

Finding Ancestors

In order to understand the origins of human traits at any level, we must have an accurate picture of our history and of the characters that distinguish it. We cannot just simply take a snapshot of humans, chimps, and other living apes and then infer how the differences among these forms are made. Each of these species has an independent lineage that reaches back as much as 6 million years or more. To get a picture of the magnitude, rate, and order of changes within or between species, we rely entirely on fossil evidence. Ever since Darwin's time, generations of paleontologists have sought to uncover the history of human origins.

The record of deep human history first began to be revealed in 1856. As workers were digging mud from a limestone cave in the Neanderthal Valley in Germany, they discovered a skull, some ribs, arm and shoulder bones, and part of a pelvis. At first, one worker took the skeleton to be that of a bear, but the brow ridge on the skull and other features convinced a local school teacher that this find might be something special—but what? It took a few years to sort out reality from various guesses.

Anatomist Hermann Schaaffhausen concluded that the bones were those of a member of an ancient race of European barbarians. A leading German pathologist pronounced that the unusual bone structures were just a consequence of rickets. Another anatomist decided that the leg bones were bent through horseback riding, and that the remains were of a Cossack soldier who had received a mortal wound in battle with Napoleon's army and crawled into the cave to die.

None of these explanations satisfied Thomas Huxley. Darwin's bulldog could not see how a dying man climbed up seventy feet inside this cave, nor how or why he buried himself without equipment or

clothes. No, Huxley concluded, this skeleton had odd, apelike characteristics. He was part of our genus *Homo*, but different. The great
geologist Charles Lyell determined that bones found nearby were those
of an extinct mammoth and woolly rhinoceros and, therefore, the
"Neanderthal" skull was of great antiquity (figure 10.1 compares *H.
neanderthalensis* and *H. sapiens* skull features).

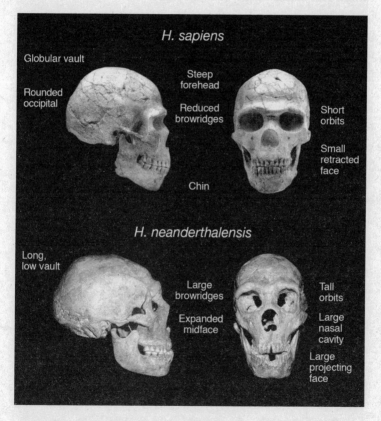

FIG. 10.1 **Comparison of *H. sapiens* and *H. neanderthalensis*
skulls.** Differences between the skulls are noted. PHOTOS COURTESY
OF DR. DANIEL LIEBERMAN, DEPARTMENT OF ANTHROPOLOGY, HARVARD
UNIVERSITY

The recognition of these bones as fossil humans could not have been more timely: the appreciation for and wider knowledge of these skeletons came on the heels of the furor over *The Origin of Species* in 1859. Although Darwin had very carefully avoided the topic in his opus, other than the one sentence "light will be shed on the origin of man and his history," the evolution of humans was, of course, the topic that aroused the most passion—both then and now.

It was Huxley who took the lead in explicit discussion of human origins. Huxley's brilliant *Evidence as to Man's Place in Nature* (1863) illustrated the state of human relations, with its frontispiece depicting the skeletons of great apes and man (figure 10.2). *The Athenaeum* magazine derided Huxley and his supporters as degraders of man's nobility who would make man "a hundred thousand years old." Ironically, this was a remarkably good guess as the earliest *H. sapiens* fossils now known are dated at about 160,000 years old.

The story has come a long way since this first Golden Age of paleoanthropology. The fossil record has continued to expand our knowl-

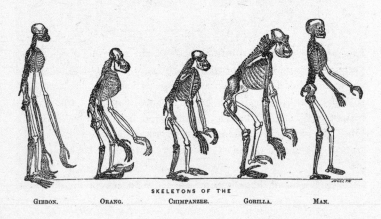

SKELETONS OF THE

GIBBON.   ORANG.   CHIMPANZEE.   GORILLA.   MAN.

FIG. 10.2 **Evolution of ape and human skeletal forms.** FROM THE FRONTISPIECE OF T. H. HUXLEY'S *EVIDENCE AS TO MAN'S PLACE IN NATURE* (1863)

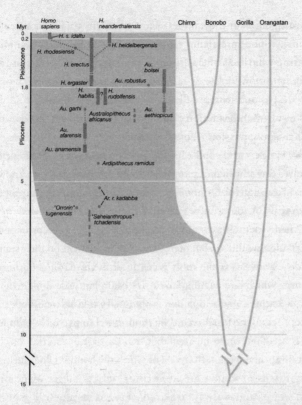

FIG. 10.3 **Hominid evolutionary tree.** The relationships between various apes and human fossil lineages are shown. This is a conservative tree that does not include all proposed species. The time span of fossil lineages is indicated by shaded bars. Note that the history of *H. sapiens* is but a small fraction of the roughly 6 million years of hominin evolution. DRAWING BY LEANNE OLDS; THANKS TO DRS. TIM WHITE AND BERNARD WOOD FOR INPUT AND ADVICE

edge, with some of the most provocative discoveries coming in just the last few years. The current spectrum of fossils informs us about three of the most crucial issues in hominid evolution (the term "hominid" refers to both humans and the African apes; "hominin" refers only to humans and our ancestors back to our separation from the apes). First, what distinguishes the hominin lineage from the apes? Second, what distinguishes modern humans (*Homo sapiens*) from earlier hominins? And third, what was the nature of the last common ancestor of hominins and chimpanzees?

Over the past two decades, the number of recognized hominin species, as well as the number of proposed species, has grown significantly. Depending upon many elements of interpretation, such as whether some fossils are variants of the same species, or "chronospecies"—a single line that evolves over time into a morphologically distinct form—we know of between fifteen and twenty hominin species dating back over the past 6 to 7 million years. A conservative picture of the hominid tree is shown in figure 10.3 (conservative in that several other fossils have been proposed to represent additional taxa, but there is not full agreement on their status). The oldest hominin is the most recently discovered, *Sahelonthropus tchadensis*, which had a chimpanzee-size brain but homininlike dental and facial features. As the hominin evolutionary tree becomes fuller and stretches back to the point where we think the chimps and human lines split, it is becoming more apparent that near the base of this tree may be a number of apelike species, from one of which the hominin line emerged.

Body fossils or crania are known for only a subset of hominin species, so we cannot always make conclusions about every aspect of anatomy that we might like. However, sufficient material is available to recognize some trends in the evolution of those hominin characters that differentiate us from other apes. The morphological or developmental characters of primary interest in human evolution are:

* Relative brain size
* Relative limb length
* Cranial size and shape
* Body and thorax shape
* Elongated thumb and shortened fingers
* Small canine teeth
* Reduced masticatory structures

* Long gestation period and life span
* Skull in upright position on vertebral column
* Reduced body hair
* Dimensions of the pelvis
* Presence of a chin
* S-shaped spine
* Brain topology.

In addition, associated anthropological evidence, such as tools, reflects the capability and behaviors of individual species, as well as the state of the evolution of certain cognitive or motor skills. Tool use was evident as early as 2.5 million years ago with *Homo habilis*.

In general, more recent species are notable for their larger body size, relatively larger brains, longer legs relative to the torso, and smaller teeth, while earlier species had smaller brains and bodies, shorter legs relative to the torso, and large teeth. The important points to keep in mind are the timescale, magnitude of character change, and the number of species over which these changes occurred. Regardless of the exact branching pattern of the hominin evolutionary tree, change was occurring over an extensive time span and many species. It is crucial to realize that our own species has been around for only a tiny fraction (about 3 percent) of the total time span of hominin evolution. Most of the physical evolution of interest predated the origin of *H. sapiens*.

Some of the major physical traits that distinguish us are not singular changes, but involve concomitant evolution of the skeleton and musculature. For example, the evolution of bipedal locomotion required changes in the vertebral column, pelvis, feet, and limb proportions and freed up the hands to evolve greater dexterity. Chimpanzees can walk on two legs when necessary, but their gait is entirely different, and they cannot extend their knee joint to straighten their leg.

The evidence for bipedalism in early hominins is derived from features of skeletal morphology. The most stunning evidence of all, though, was discovered in 1976 around the Laetoli archeological site in

Tanzania. Paleoanthropologist Andrew Hill was engaging in some typical primate behavior, tossing elephant dung at a colleague, when he stumbled upon sets of hominin footprints that trailed for about eighty feet through a volcanic ash bed (figure 10.4). These astounding prints were made by at least two individuals, one large, one small, who were walking through a fresh ash fall 3.6 million years ago. These prints were then covered until Hill's discovery and Mary Leakey's field team excavated and studied the site. The only known hominin species of that age in that location was *Australopithecus afarensis*, a small-brained, upright-walking species first made famous by the "Lucy" skeleton discovered in Ethiopia by Donald Johanson.

Fig. 10.4 **Ancient hominin footprints.** These footprints in an ancient ash bed, inferred to be those of an adult and juvenile *Australopithecus afarensis*, were discovered in 1976 at Laetoli, Tanzania. PHOTOS COURTESY OF PETER JONES AND TIM WHITE, UNIVERSITY OF CALIFORNIA—BERKELEY

While bipedalism and its associated features evolved early in our lineage, our large brains did not. Australopithecines such as *Au. afarensis* and *Au. africanus* had brains about 450–500 cm^3 in volume, not much larger that that of a chimpanzee (about 400 cm^3). Brain and body size increased dramatically in the genus *Homo* in the past 2 million years (figure 10.5), but again this was not a simple, steady increase. Rather, there appears to have been a burst in absolute brain size by the early

Brain and Body Size Evolution in Homonins[a,b]

Species	Estimated age (MYA)	Body size (kg)	Brain size (cm3)
Homo sapiens	0–0.2	53	1355
H. neanderthalenis	0.03–0.3	76	1512
H. heidelbergensis	~0.3–0.4	62	1198
H. erectus	0.2–1.9	57	1016
H. ergaster	1.5–1.9	58	854
H. habilis	1.6–2.3	34	552
Paranthropus boisei	1.2–2.2	44	510
Au. africanus	2.6–3.0	36	457
Au. afarensis	3.0–3.6	NA	NA
Au. animensis	3.5–4.1	NA	NA
Ardipithecis ramidus kadabba	5.2–5.8	NA	NA
Sahelanthropus tchadensis	6–7[3]	NA	~320–380

[a] Not a complete list of all unrecognized or proposed species; some dates and relationships between species are unresolved.

[b] See Sources (page 307) for reference material.

FIG. 10.5 There is a broad trend toward an increase in body and brain size from older to more recent species. Body fossils or complete skulls are not available (NA) for all species.

Pleistocene (1.8 million years ago) and another in the middle Pleistocene (600,000–150,000 years ago), separated by a period of about 1 million years of relative stasis.

Why did our brains get so much bigger during these periods? There are many theories. I'll mention just one, the adaptation to climatic change, because I think it reflects a view that is becoming more widely accepted about the role of external forces in driving the pace of evolution. About 2.3 million years ago, there began a global shift toward a cooler and drier climate. This caused the forests of Africa to shrink and to be replaced with drier savannah. While the great apes stayed in more stable rain-forest habitats, hominins adapted to more variable habitats. After a period of relative stability, in the last 700,000 years the Earth's climate has, on average, been colder than during any other period since the extinction of the dinosaurs 65 million years ago. Abrupt fluctuations in temperature have occurred many times, with some major shifts taking place in the course of just several years. The changing climate and its effects on food availability, water, hunting, and migration may have selected for hominins better adapted to such constantly changing conditions. Under the changing climate, brain size roughly doubled in a million years, encompassing perhaps 50,000 hominin generations. This is impressive, but far from instantaneous.

It is fun to point out that body and brain size were even greater in Neanderthals than in modern humans. We have no obvious physical indication of why we succeeded and our cousin died out about 30,000 years ago without leaving descendants. Our line and the Neanderthal lineage split off from each other well before the origin of *H. sapiens*, around 500,000 years ago. *H. neanderthalensis* did not contribute to the *H. sapiens* gene pool. This was demonstrated conclusively in a remarkable study, one of the really great contributions of genetics to paleoanthropology. Svante Paabo and his colleagues, then at the University of Munich, managed to sequence DNA extracted from a bone of a Neanderthal specimen, and this sequence proved that Neanderthals are a dead twig on the human evolutionary tree.

H. sapiens and Neanderthals did coexist: several sites have been uncovered that reveal the presence of both species at the same time. Both species used tools, made fire, and had other signs of culture, language, and self-awareness, but only one prevailed. Whatever intellectual advantage modern humans might have held over Neanderthals as they took over their range, this would most likely have been subtle in terms of neuroanatomy and will be difficult to ascertain. However, the bigger picture of hominin brain development and evolution, relative to that of the great apes, appears more tractable.

The Making of a Beautiful Mind

The marked increase in brain size in more recent hominins is only a crude measure of a potential increase in cognitive capabilities. Absolute brain size is not necessarily an indicator of greater power. What is more telling is the relative increase in brain size compared with body mass. The brain is a very expensive organ in terms of energy consumption, drawing up to 25 percent of an adult human's energy (and 60 percent of an infant's). The relative increase in hominin brain size in the Pleistocene is a marked deviation from typical mammalian and primate ratios. While whales and elephants have much larger brains than we do, our brains are, as a percentage of body weight, 15–20 times larger than those of these mammals. The challenge for neuroanatomists has been to identify what aspects of brain increase are most meaningful in terms of human capabilities.

The magnitude of this challenge was captured by Emerson Pugh, an IBM computer research scientist, who wrote, "If the human brain was so simple that we could understand it, we would be so simple that we couldn't." Understanding the brain and understanding the biological basis of behavior are two of the great frontiers yet to be conquered in biology.

The roles of certain areas of the brain in visual, motor, and cognitive functions have been well studied in mammals and primates, including humans. The top portion of our brains, the cerebral cortex, is a sheet of neural tissue that covers most of the brain. Part of this sheet, the six-layered neocortex, is a structure found only in mammals. In humans, the cortex is arbitrarily divided into several lobes whose boundaries are defined by particular grooves and bumps on the brain surface. Neurobiologists have been particularly successful in localizing functions to lobes (figure 10.6). This includes the frontal lobe, which is involved in thinking, planning, and emotion; the parietal lobe, which is involved in the sensation of pain, touch, taste, temperature, and pressure, as well as in mathematics and logic; the temporal lobe, which is primarily involved with hearing, as well as memory and the processing of emotions; the occipital lobe, which is involved in the processing of visual information; and the limbic lobe, which is involved in emotional and sexual behavior, and in memory processing.

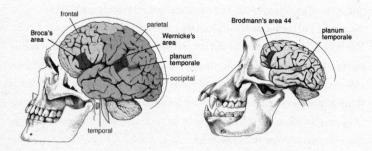

FIG. 10.6 **Physical landmarks in human and chimpanzee brains.** Broca's area and Wernicke's area in the human planum temporale are associated with speech functions. Anatomical features associated with these structures have been reported in chimpanzees. DRAWING BY LEANNE OLDS

One of the first areas of the cortex to be functionally identified was mapped by Paul Broca, who in 1861 examined the brain of a stroke patient who could utter only a single word, "tan." Broca found a lesion in the frontal lobe of the brain and concluded this was a speech area. His observations have since been supported by many kinds of evidence, including imaging of normal brains of individuals when speaking. Ever since Broca's era, comparative neuroanatomists have sought to identify areas that might be central to the evolution of human talents. The major point that can be drawn from comparisons of brain anatomy is reminiscent of the stories I have told about other inventions, such as butterfly wing spots, spider spinnerets, and the insect wing—namely, that the current form of the structure owes itself to many inventions that long predated what we see now. Mammalian brains are distinct from what preceded them, early primate brains are a further elaboration upon the mammalian foundation, and ape and human brain evolution was superimposed upon the advanced state of the primate condition.

A key early invention was certainly that of the neocortex in mammals. Not only did this add processing power to the brain, but it opened the way to evolutionary specialization in particular functional subsystems. Changes in brain size among mammals are not simply a matter of enlarging or reducing all parts of the brain proportionately. Rather, brain evolution exhibits a "mosaic" pattern, with certain parts of the brain changing in concert with another, but independently of other parts. For example, the tenrec (an insectivore, a small bug-eating mammal) has a non-neocortical brain volume that is greater than that of the marmoset (a primate), but the marmoset's neocortex is almost ten times larger (figure 10.7). In primates overall, the neocortex has been expanded such that it is, on average, about 2.3 times larger than in nonprimates of similar body weight. Within primates, shifts from a dependence on a sense of smell to a greater reliance on vision are associated with relative shifts in the sizes of cortical areas involved in each task.

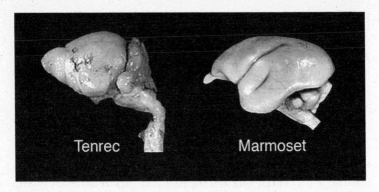

Tenrec   Marmoset

FIG. 10.7 **Evolution of mammalian brain areas.** The tenrec, an insect-eating mammal, has a much smaller cerebral cortex than the marmoset, a primate. Relative shifts in the size of brain regions are a common feature of specializations. PHOTOS COURTESY OF CAROL DIZACK AND WALLY WALKER, THE WISCONSIN COMPARATIVE MAMMALIAN BRAIN COLLECTION, UNIVERSITY OF WISCONSIN

In addition to shifts in proportion, new centers have evolved. One area of the primate brain that appears to be novel is a center for coordination of visually guided motor activities. Reaching for, grasping, and manipulating objects are obviously important for primate lifestyles. There is a region called the ventral premotor area that is activated during visually guided movements and, very interestingly, also when monkeys observe these tasks being performed. This suggests that this primate motor area may be critical to learning through visual observation.

Because speech and language have played such an enormous part in our evolution, the origin of these capacities has been of immense interest. Broca's area in the human brain is located in the primate premotor area and may be a specialization for speech and language. The burning issue has been whether brain areas for these activities are unique to humans. One of the gross anatomical features associated with Broca's area is that this region is larger in the left hemisphere of the brain than in the right. We know that the left hemisphere of the brain dominates speech production, so this asymmetry in Broca's area has been proposed to reflect the specialization of the left hemisphere. The left hemisphere also controls right-handedness, and hand gestures are also part of our communication process. A second language area, called Wernicke's area, is in the temporal lobe (figure 10.6). A site within this area, the planum temporale, is implicated in spoken and gestural communication and in musical talent, both of which are also left-hemisphere dominant. In most humans, there is also a left-right hemisphere anatomical asymmetry in this area where a particular fissure extends farther back in the left hemisphere than the right.

Evidence for these anatomical asymmetries has been reported in great apes. This would suggest that anatomical areas that have become specialized in humans were also defined in the common ancestor of humans and great apes. There is also some evidence that communication in captive apes is left-hemisphere dominated, so that would support the inference that the anatomical architecture for communication long predated hominins. However, more recent studies on a larger number of specimens are not supportive.

There is now clear evidence in humans that these anatomical asymmetries are not required templates for speech production or handedness. In 1 out of about 10,000 humans, the normal left-right asymmetry of the distribution of internal organs is reversed (this is called "situs inversus"), but these individuals are generally functionally normal. Recent imaging studies of the brains of situs inversus individuals reveal that the left-right asymmetries in the frontal lobe and planum temporale are also reversed. However, these individuals still have left-hemisphere dominance in speech production and are generally right-handed. These observations show that these two long recognized anatomical asymmetries in the human brain are not necessarily required for the development of speech and language function.

Gross and detailed comparisons of human and ape brains have been undertaken to search for other areas that might account for our functional aptitudes. There is a long-standing idea that areas of the brain involved in planning, organizational behavior, personality, and other "higher" cognitive processes might differ. These properties map to areas of the frontal cortex, which is larger in humans than in chimps, but not disproportionately so. Could it be then that what makes us different is more subtle? Probably. The stuff of our evolution is more likely to be found in the "microanatomy" of our brain, including the interconnections between cortical regions, the architecture of local wiring circuits, or the arrangement of neurons in the cortex. For example, the dimensions of vertical columns of neurons in the planum temporale do differ between chimps and humans. Evolutionary tinkering in the number, arrangement, and connectivity of neurons formed during development in specialized areas of our ancestors' brains was the most likely path to the origins of our capabilities. Neurobiologists are in hot pursuit of potential fine-scale differences among apes and human brains using a host of high-resolution technologies.

The Mosaic Evolution of Human Form and Development

The physical differences between the forms of modern humans, earlier hominins, and the great apes are the products of evolutionary changes in development. In order to understand the nature of those changes, detailed study has been made of the rates of growth and maturation of humans and chimps, and some deductions have also been made from fossil material.

One of the long appreciated, fundamental differences in chimpanzee and human development is the relative rate of skull growth and maturation. Human babies have less mature skulls in terms of their shape than do young chimpanzees, even though human skulls and brains are much larger. In humans, the maturation of the skull is slowed dramatically compared with the chimp, which allows for its greater initial size.

Chimpanzee and human skulls eventually grow to the same size, but attain very different face sizes and brain case volumes. The relative shift in skull maturation rates indicates that the timing of similar developmental processes has been shifted.

Evolutionary shifts in the timing of other developmental processes have also been revealed from the study of hominin fossils. Paleontologists can tell from enamel patterns on fossil teeth that tooth formation times were shorter in Australopithecines and early *Homo* species than in modern humans. The stages of dental development are reliable indicators of stages of juvenile development and the relative age of sexual maturation. The fossil record reveals that these aspects of modern human biology appeared later than other changes such as those of brain size and body proportions. In contrast, all of the skeletal changes associated with our bipedal posture are due to structural changes in bones and musculature and were attained earlier and independently of the slower maturation of the skull. Overall then, the picture of hominin evolution is a mosaic pattern, with different traits appearing at different times and evolving at different rates in hominin history.

The importance of this mosaic pattern to an Evo Devo perspective on human evolution is that it tells us that the development of different structures was evolving in a patchy, nonlinear way over a long course of time. The fossil record dispels any notion of a sudden, instantaneous change in human form. Rather, our history involved quantitative shifts—in brain size, body proportion, skull size, gestation time, juvenile development, and more—assimilated over tens of thousands of generations. Furthermore, the rates of change in human characters were not at all exceptional with respect to what was also transpiring in other mammals during the time period of human evolution. For instance, fossil horses show similar rates of change in body size and other characters.

The body of evidence tells us that the evolution of human form was not special or atypical of other animals. We should expect then that what we know about the evolution of animal form, in general, applies as well to humans. Indeed, our extremely close genetic relationship to chimpanzees, as well as the genetic similarity of primates to other mammals, underscores a now familiar theme. The sets of genes for making these animals and humans are very similar; the differences in form between them, both great and small, must lie in how they are used—or, as we will see in one case, not used.

The 98.8 Percent Paradox and the Making of *Homo sapiens*

The ultimate cause of developmental and physical change in the evolution of humans is genetic. Somewhere in our DNA reside the differences between ourselves and apes, and earlier hominins. The critical questions then are:

* How many significant differences are there?
* Where are they?
* How have they contributed to differences in form?

The good news is that we now have complete genome sequences for a human, a chimp, and a mouse.

The bad news comes from doing a little arithmetic.

The DNA sequence of a human contains about 3 billion base pairs. In chimp DNA, about 98.8 percent of these bases are identical. That is a difference of just 1.2 percent, the smallest fraction of DNA sequence differences between ourselves and any other animal on the planet. However, that 1.2 percent difference translates to *36 million* different base pairs. Because humans and chimps diverged from a common ancestor about 6 million years ago, we can assume that one-half of these differences are chimp-specific (occurred in the chimp line) and one-half are human-specific (occurred in our lineage). That still leaves

18 million changes in our line since our last common ancestor. (I am simplifying the numbers here a bit for discussion purposes. I am not counting the deletion or insertion of bases, or the gain or loss of larger DNA elements.)

Do all of these changes matter? Or, are some just noise? How can we decide which of these 18 million differences contributed to evolution?

We do know that not all mutations in a gene are meaningful. Because the genetic code is redundant, some bases can change without altering a protein. These "silent" substitutions accumulate as a function of time because there is little or no selective pressure to eliminate them. In addition, because only about 5 percent of our DNA is involved in coding or regulatory functions, mutations occurring elsewhere in the vast expanse of human DNA sequence are of no or little consequence. Furthermore, an additional fact to consider is that any two unrelated humans will differ, on average, at about 3 million bases. While that seems to be a large number in absolute terms, it is only 0.1 percent of all DNA bases; and, despite these differences, we clearly belong to one species. This tells us that millions of differences may be of no consequence whatsoever. So, in fact, nobody knows how many changes shaped human form. My guess would be somewhere on the order of a few thousand. The challenge now is to find those differences that do matter.

Before I analyze chimp-human differences further, I think that the paradox, and its general solution, becomes clearer when we compare the human genome with that of another mammal, the mouse. Mice are rodents, and the rodent and primate lines separated a long time ago, probably on the order of about 75 million years ago. Mice are small-brained; they possess a neocortex but it is much smaller relative to that of primates, and, of course, minuscule in comparison to ours. Yet, comparison of mouse and human genomes reveals that greater than 99 percent of all genes in the human have a mouse counterpart, and vice versa. In fact, 96 percent of all genes in the human are found in the exact same relative order in human chromosomes as in the mouse chromosomes. This is a remarkable degree of similarity. These figures tell us that in the course of 75 million years of mammalian evolution, and at least 55 million years of primate evolution, our genome and that of a rodent contain essentially the same genes in mostly the same organization. Differences in gene number and organization have not played much, if any, role in the origin of humans or primates.

If not the number or organization of genes, what else could explain the enormous differences between mice and humans? The sequences of proteins encoded by mouse and human genes do differ, by about 30 percent on average. But, based on what we have seen so far, are the differences in protein sequences likely to account for most changes in form?

Generally speaking, I don't think so. I make this argument based more on what we know from other species rather than on direct experimental data on humans, but I think this conclusion is inescapable based on several lines of evidence. First, most proteins in the body do not affect form—they carry out other roles in physiology. There may be some interesting differences in proteins involved in physiology, such as the sense of smell, immunity, or reproduction, but these do not affect the way mice or humans appear. Second, the tool kit proteins are a small fraction of all the proteins in the body, and we have seen that because each protein usually has many jobs in development, they are even less likely to change in meaningful ways (because mutations would usually affect all functions, not just one). Rather, as we have seen in previous chapters, changes in genetic switches account for many differences in animal form. Because human evolution is largely a matter of the evolution of the size, shape, and fine-scale anatomy of structures, and of timing in development, it is only logical that switch evolution would be important in the evolution of humans as well. Everything in our bodies is a variation on a mammalian or primate template. Thus, I believe that the weight of genetic evidence is telling us that the evolution of primates, great apes, and humans is due to changes more in the control of genes than in the proteins the genes encode.

I am not the first to reach this conclusion. In a classic study three decades ago, Mary Claire King and Allan Wilson showed that the sequences of chimp and human proteins were nearly identical and drew the conclusion that evolutionary differences were due to changes in gene regulation. A host of eminent biologists in the 1960s and 1970s—including Linus Pauling, Emile Zuckerkandl, Eric Davidson, Roy Britten, and François Jacob—also deduced the same. However, at the time, we knew nothing at all about the logic and function of genetic switches in animals, nor of even a single gene that controlled development. The weight of evidence from Evo Devo and comparative genomics tells us that these earlier deductions were on the right track.

However, despite their importance, it is much more difficult to study human gene switches than those of other species (because we can't study their function in living human embryos). This makes the identification of evolutionary changes in human switches very challenging. While various kinds of efforts are under way, it has been easier thus far to spot differences in protein-coding sequences that may be responsible for or associated with aspects of human evolution. I will focus on two genes that have been implicated in human evolution. Their stories illustrate the kind of detective work that is needed to implicate particular genes with the evolution of human features. These examples should be taken as illustrative of how such associations are made, as the first stars to be seen through new genetic telescopes. They are not necessarily, or even likely to be, the most important or exclusive genetic causes of the evolution of these traits.

Evolution of Human Jaw Muscles

Among the traits that distinguish us from other apes, or earlier hominins such as the Australopithecines, is the reduced size of our jaw muscles. Living primates, such as the macaque or gorilla, have large, powerful jaw muscles for breaking down food. One muscle that ele-

macaque   gorilla   human

FIG. 10.8 **Evolution of jaw musculature in primates.** Macaques and gorillas have a large temporal region to which the temporalis muscle attaches. This large region is necessary to generate sufficient force for the large jaw and chewing pressure of these animals. In humans, the temporalis is much reduced—this feature is correlated with at least one mutation in a muscle fiber protein. COURTESY OF DR. HANSELL STEDMAN; FROM *NATURE* 428 (2004): 415; REPRINTED BY PERMISSION

vates the mandible, the temporalis, is attached over most of the temporal region of the skulls of living primates, but is much reduced in proportion in humans (figure 10.8).

One genetic clue to the origin of the shift in jaw muscle size has been uncovered by Hansell Stedman and colleagues at the University of Pennsylvania. They noticed that the human gene encoding a particular protein called myosin heavy chain 16 (MYH16 for short) had a mutation in it that disrupted most of the protein sequence. Myosin heavy chains play a crucial role as parts of the fibers within muscles that generate force by contraction. When these proteins are absent or altered, the fibers and muscles are usually reduced in size.

MYH16 is a specialized myosin found in only a subset of muscles. In the macaque, MYH16 is made in the temporalis muscle and a second muscle nearby, but not in others. The human *MYH16* gene is expressed in the human temporalis muscle, but the mutation in the gene has inactivated the protein's function. The muscle fibers in the human temporalis are only about one-eighth the size of those of the macaque. This genetic and anatomical evidence suggests that the inactivation of the MYH16 protein is associated somehow with the reduction of the temporalis muscle sometime in hominid evolution.

When might this genetic change have occurred? It definitely occurred after the split of the human and chimp lineages because chimps (as well as other apes and monkeys) have an intact MYH16 gene that encodes a full-size MYH16 protein. Based upon the number of changes in the human gene relative to other species, the Pennsylvania group has estimated that the inactivating mutation occurred somewhere between 2.1 and 2.7 million years ago. This is tantalizingly close to the period of the origin of the genus Homo.

The significance of the evolutionary reduction in jaw musculature extends beyond how hominins chewed their food. Muscle anatomy has a large influence on bone growth, and experimental studies have shown that jaw muscle growth has a significant impact on the size and shape of the craniofacial skeleton. Reduction in the jaw musculature, and the force imposed on the mandible, would reduce the stress on bones in the skull. This could have allowed the braincase to become thinner and larger. Thus, the expansion in brain size that took hold in early Homo may have been enabled, to some degree, by changes in jaw musculature and related skull features. Furthermore, the reduction in jaw musculature may have facilitated the eventual evolution of finer control of the mandible, as is required for speech.

All of these connections and associations are intriguing, but we must be cautious not to attribute all of this anatomical change to a single mutation. While the inactivation of the previously functional MYH16 gene is certainly a noteworthy occurrence, we cannot say whether this inactivating mutation was the initial genetic change toward reduction of the temporalis, or one of many sequential or parallel changes, or the last change that occurred once the role of the MYH16 protein in the temporalis became dispensable. For reasons I will explain shortly, there is no reason to assert that it was the single, critical evolutionary trigger. This will always be difficult to ascertain for any gene implicated in human evolution, such as, for example, the recent discovery of a gene implicated in the evolution of human speech.

Evolution of a Gene Affecting Speech

One advantage that searchers for human genes of potential evolutionary interest have is that there is a lot of us, about 6 billion now, and when some function isn't quite right, humans show up at medical clinics. This allows even very rare mutations to be discovered that might arise only once in 1 billion individuals. One striking example of such a very rare and informative mutation was detected in a small family in which members of three generations exhibited a severe speech and language impairment. What is most interesting about the affected individuals is that their impairment is not due to some muscular problem with the production of speech; rather, they have some deficits in the neural circuitry that affects language processes. State-of-the-art imaging techniques have revealed that affected individuals have some detectable abnormalities in several brain areas. Furthermore, magnetic resonance imaging of affected individuals during the performance of silent (thought) and spoken tasks reveals underactivity in Broca's area and a few other language-related areas. The patients appear to have a deficit in a neural network involved in the learning and/or planning and execution of speech sequences.

The gene that is mutated in this family has been identified and is called *FOXP2*. The FOXP2 protein is a transcription factor, a tool kit protein that binds to DNA and regulates the expression of other genes. This mutation changes one amino acid in FOXP2, and this one change appears to knock out the FOXP2 protein function. Because these patients also carry one copy of the normal *FOXP2* gene, they still have some FOXP2 function. The speech and language impairment is due to a reduction in the total amount of functional FOXP2 protein, not a complete loss. Perhaps the first question that springs to mind about *FOXP2* is whether it is a novel, uniquely human gene.

I hope that all I have said for the past several chapters has prepared you to guess the answer to that question. No, *FOXP2* is not at all unique to humans. The gene has been identified in a bunch of primates, rodents, and a bird. This distribution is typical of human tool kit genes, in that most, if not all, have counterparts in other species. In fact, the human FOXP2 protein differs at only 4 out of 716 positions from that of the mouse, at 3 positions from the FOXP2 of the orangutan, and at just 2 positions from those of the gorilla and chimpanzee. This is a smaller amount of change in sequence than most proteins show, indicating that there has been a lot of pressure to conserve the FOXP2 protein sequence throughout mammalian evolution.

Did the evolution of the *FOXP2* gene play a role in the origin of speech and language? This is a trickier question to answer; the changes in *FOXP2* are more subtle than the inactivating mutation in *MYH16*. Another way to test whether a gene might have played a recent role in evolution is to look for signs of what is called a "selective sweep." The action of natural selection can leave a trail of evidence in the form of the pattern of DNA sequence variation that arises after a favorable mutation is selected for. Variation in a length of DNA sequence accumulates as a function of time unless or until selection acts to favor a particular variant. Selection for a variant causes a "sweep" that reduces overall variation. From the pattern of reduced variation at a gene relative to its neighbors, geneticists can tell if a gene has experienced a selective sweep. The signal of a selective sweep at the human *FOXP2* locus is one of the strongest at any human gene. This is a good indication that for some period of the past 200,000 years, during the evolution of our species, mutations in the *FOXP2* gene were favored and spread throughout *H. sapiens*.

What changes in *FOXP2* might have contributed to the evolution of speech? There are just two coding differences between the human and chimp proteins. While it is possible that these could be responsible, there are hundreds more differences in noncoding DNA around *FOXP2*, in switches and regions that affect the place and amount of *FOXP2* expression. It is very difficult to pinpoint the changes that might have been meaningful to human evolution with current technology. My money is on the noncoding regions because tinkering with switches of the *FOXP2* gene would allow fine tuning of *FOXP2* expression in the formation of neural networks. It is known that *FOXP2* is expressed in many sites in the developing human brain. It is also expressed in the counterparts of these regions in the mouse, so FOXP2 would appear to have a widespread role in mammalian brain development. It is not clear yet precisely what FOXP2 does in development, but it is likely that it affects how subregions of the brain form and connect with other parts. Because, as I have said again and again, it is difficult to change a tool kit protein in such a way that only one or a subset of functions is affected, I suspect that evolution in the switches controlling the *FOXP2* gene has enabled the evolution of fine-scale differences in individual brain regions.

The Complex and Subtle Genetic Basis of Human Evolution

The discoveries of *FOXP2* and *MYH16* have generated a great deal of excitement in scientific and medical circles, as well as in the general press. But are they the whole story of the development and evolution of jaw musculature and craniofacial form, or of speech and language? Not at all. They are just the beginning. In order to put the discovery and role of *FOXP2* and *MYH16* in context, we have to shed a long-standing tendency in many circles, including scientists as well as the general press, to imagine evolution occurring in one leap through the occurrence of a single dramatic mutation. Such ideas have been for-warded for the origin of speech and language and other complex human traits and are often coupled to the idea that the evolution of some characteristic was "rapid." But we have seen that brain size, skeletal anatomy, dental development, skull shape, and other features evolved over many tens of thousands of generations or more. There is no need to invoke single dramatic mutations as causes of great leaps in form and function or as explanation for the origins of human traits. Nor is there any scientific foundation for doing so.

The term "genetic architecture" has been coined to refer to the num-ber of genes and the relative effect of individual genes contributing to the evolution of a particular trait. Decades of work on quantitative characters, such as body size, or the number of some particular struc-ture has shown that variation within species, or differences between species, are often due to many genetic differences that are individually responsible for relatively small effects. They suggest that evolutionary shifts in characters occur in small increments, via changes in poten-tially many genes. The genetic architecture of human trait evolution should be no different and, in fact, studies of human variation suggest that many genes contribute to differences in height, body size, and other quantitative characters. We do not know whether the inactiva-tion of *MYH16* was an early step in the evolution of the temporalis muscle, or a very late step after the *MYH16* function became irrele-vant. It is very likely that changes in other genes contributed to the reduction in jaw musculature over an extended time period. Likewise, *FOXP2* is certain to be just one part of the story of the evolution of speech. We should expect that selection for evolutionary changes at other genes—or, more specifically, in their genetic switches—also con-tributed to the evolution of this human talent. We know about *FOXP2* because of a lucky strike, a one-in-a-billion mutation that happened to be clinically detectable when present in just one copy of the gene. We know about *MYH16* because we can easily tell that this gene was inac-tivated. There are many more genes to discover and study in stitching together the history of human evolution, some of which will have much more subtle effects and histories than these two genes.

Because there are more stories like that of *FOXP2* and *MYH16* to come, we should resist the natural tendency to treat new discoveries— of a fossil, a brain landmark, or of a particular gene—as "the" solution to the puzzle of human evolution. Rather, most discoveries are individual pieces of a more complex mosaic. Paleoanthropology now recognizes a complex pattern of hominin evolution, with more species than earlier views, as well as numerous dead twigs rather than a single straight line leading from a distant ancestor to modern humans. In fact, as more fossils are discovered that cluster near the fork between the human and chimp lineages, any claim to finding "the" ancestor should be viewed with suspicion. Similarly, comparative neurobiology must now search for more subtle explanations of human capabilities, as the most obvious anatomical landmarks in the brain appear to have deeper origins than first thought and do not strictly account for human behaviors. Similarly, it is very unlikely that the evolution of any traits that define us—bipedalism, skeletal form, craniofacial form, brain size, or speech—was the result of selection on just a few major genes. *FOXP2* and *MYH16* are the first pieces in the puzzle to be identified, but we have no reason to believe they are the biggest or most important ones. Rather, the more likely picture is that hominin evolution was forged by selection for variants of many genes, responsible for small increments of differences in size, shape, and tissue composition, over sustained intervals of many thousands of generations.

I mention this caveat to oversimplifying new discoveries not to diminish the excitement they warrant, but because of the larger issues at stake in deciphering the material basis of human evolution. Evolutionary biology has faced resistance since birth and it has been difficult enough for basic concepts drawn from rock-solid data on finches, moths, or fruit flies to gain acceptance. Some claims about human evolution are certain to require revision as more data become available, as has occurred continually in paleoanthropology for the past century. Opponents of evolutionary sciences always seek to exploit even appropriately cautious statements by scientists as evidence of doubt or uncertainty and grounds for not teaching evolutionary principles. Simplification may indeed be necessary for news articles, but it can distort the more complex and subtle realities of evolutionary patterns and mechanisms.

The discoveries of Evo Devo are illuminating the evolutionary process and particular evolutionary events in powerful new ways. Evo Devo has expanded the foundations of evolutionary biology, changed the way we think, and provided a new opportunity to change the way evolutionary biology is reported, taught, and discussed. In my concluding chapter, I will discuss the place of Evo Devo in an overall evolutionary synthesis and the role it must play in the teaching of evolutionary biology, and in the vanguard of the perennial social controversy over evolution.

The geometric beauty and diversity of seashells. JAMIE CARROLL

11

Endless Forms Most Beautiful

> Observe always that everything is the result of a change, and get used to thinking that there is nothing Nature loves so well as to change existing forms and to make new ones like them.
>
> —Emperor Marcus Aurelius Antoninus

DARWIN CLOSED THE first edition of *The Origin of Species* with what has become perhaps the most widely quoted passage in all of biology:

There is a grandeur in this view of life, with its several powers, having been originally breathed into a few forms or into one: and that whilst this planet has gone cycling on according to the fixed law of gravity, from so simple a beginning endless forms most beautiful and most wonderful have been, and are being, evolved.

It took him some twenty years to arrive at this phrasing. In earlier general sketches of his ideas, completed in 1842 and 1844, but never published, this passage was longer and notably different. The 1842 version read:

There is a simple grandeur in this view of life with its powers of growth, assimilation, and reproduction, being originally breathed into matter under one or a few forms, and that whilst this our planet has gone circling on according to fixed laws, and land and water, in a cycle of changes, have gone on replacing each other, that from so simple an origin, through the process of gradual selection of infinitesimal changes, endless forms most beautiful and most wonderful have been evolved.

In 1844, Darwin tinkered with the choice of a few words, but the major changes came with the preparation of *The Origin of Species* for publication in 1859. Darwin removed the reference to the "process of gradual selection of infinitesimal changes" and condensed other phrases in ways that simplified the text and gave to it a poetic rhythm.

I have chosen four words that remained completely untouched throughout all versions and editions, "endless forms most beautiful," as the inspiration for this book and the theme of this concluding chapter. This phrase captures the essence of the new science of Evo Devo. Here, I will discuss how the discoveries and perspectives of Evo Devo enlarge the grandeur of the evolutionary view of life, enrich our understanding of how Darwin's endless forms have been and are being evolved, and have expanded and deepened the foundation of evolutionary thought.

My publisher informs me that I will not have twenty years to find just the right words, nor could I hope to capture the essence of this new discipline in such masterly prose as Darwin's; nevertheless I will attempt to articulate four main points about the impact and importance of Evo Devo by way of summary and conclusion.

First, I assert that Evo Devo constitutes the third major act in a continuing evolutionary synthesis. Evo Devo has not just provided a critical missing piece of the Modern Synthesis—embryology—and integrated it with molecular genetics and traditional elements such as paleontology. The wholly unexpected nature of some of its key discoveries and the unprecedented quality and depth of evidence it has provided toward settling previously unresolved questions bestow it with a revolutionary character.

Second, Evo Devo provides a new means of teaching evolutionary principles in a more effective framework. By focusing on the drama of the evolution of form, and illustrating how changes in development and genes are the basis of evolution, the deep principles underlying the unity and diversity of life emerge. Furthermore, the visible forms of gene expression patterns in embryos and the concrete inventories of tool kit gene sets in different species provide more effective ways of illustrating evolutionary concepts than previous, more abstract approaches.

Third, because Evo Devo reveals and illustrates the evolutionary process and principles in such tangible ways, it has a critical role to play in the forefront of the societal struggle over the teaching of evolutionary biology.

And finally, the importance of evolutionary biology is far more than mere philosophy. The fate of the endless forms of Nature, including humans, depends on a broader understanding of human impacts on evolution.

Evo Devo as a Cornerstone of a More Modern Synthesis

> Embryology is to me by far the strongest single
> class of facts in favor of change of forms, and not
> one, I think, of my reviewers has alluded to this.

> —Charles Darwin, letter to Asa Gray,
> September 10, 1860

The above quote reflects that embryology has always been an integral component of the evidence for evolution and the principle of common descent. The challenge for more than 100 years after Darwin was to explain *how* embryos—and thus the adult forms they produce—change. The Modern Synthesis added genetics to the evolutionary edifice, but geneticists of that era were largely restricted to studying small variations within species and did not know the chemical nature of the gene (i.e., DNA), let alone how genes affected form. The major accomplishment of the Modern Synthesis was the reconciliation between paleontological views of so-called macroevolution, or evolution above the species level, with genetic views of microevolution, the variation detectable within species. The synthesis asserted that large-scale changes in form as viewed in the fossil record could be explained by natural selection acting over long periods of time upon small genetic changes that produce variation within species. This was an extrapolation supported by a consensus opinion, but no one knew whether the genetic mechanisms governing large-scale change were the same or might be different than those affecting variation within species. Nothing was known about how genes affect form, which genes affect the evolution of form, or what kinds of changes in genes were responsible for evolution. Furthermore, once the structures of DNA and proteins were understood, the prevailing view of the architects and adherents of Modern Synthesis was that the process of random mutation and selection would so alter DNA and protein sequences that only closely related species would bear homologous genes.

Virtually everything I have described in chapters 3–10 has been discovered in the past twenty years. The insights provided by these discoveries have not just filled enormous gaps in our grasp of evolutionary processes, but they have also forced biologists to rethink completely their picture of how forms evolve. You have read detailed evidence of how differences in the way ancient tool kit genes are used has shaped the evolution of animal forms from Urbilateria to *Homo sapiens*. Here, I will summarize how major evolutionary ideas have been expanded, illuminated, or reconsidered based on this new body of evidence.

On Descent with Modification

THE TOOLS FOR MAKING THE KINGDOM ARE ANCIENT

The first and still perhaps the most stunning discovery of Evo Devo is the ancient origin of the genes for building all sorts of animals (chapters 3 and 6). The fact that such different forms of animals are shaped by very similar sets of tool kit proteins was entirely unanticipated. The ramifications of these revolutionary findings are powerful and manifold.

First of all, this is entirely new and profound evidence for one of Darwin's most important ideas—the descent of all forms from one (or a few) common ancestor. The shared genetic tool kit for development reveals deep connections between animal groups that were not at all appreciated from their dramatically different morphologies.

Second, the discovery that organs and structures that were long viewed as independent analogous inventions of different animals, such as eyes, hearts, and limbs, have common genetic ingredients controlling their formation has forced a complete change in our picture of how complex structures arise. Rather than being invented repeatedly from scratch, each eye, limb, or heart has evolved by modification of some ancient regulatory networks under the command of the same master gene or genes (chapter 3). Parts of these networks trace back to the last common ancestor of bilaterians (Urbilateria), and earlier forms (chapter 6).

Third, the deep history of the tool kit reveals that the invention of these genes was not the trigger of evolution. The bilaterian tool kit predated the Cambrian (chapter 6), the mammalian tool kit predated the rapid diversification of mammals in the Tertiary period, and the human tool kit long predated apes and other primates (chapter 10). It is clear that genes per se were not "drivers" of evolution. The genetic tool kit represents possibility—realization of its potential is ecologically driven.

On Complexity and Diversity

Large-scale Trends in Animal Design and Evolution Have a Common Basis, and Are Enabled by the Properties of the "Dark Matter" of the Genome

I have focused much attention in this book on the modular construction of animals from serially reiterated parts and the trend in evolution toward the greater specialization of these parts (chapter 1). Modularity is key to the building of complexity and the evolution of diversity. Complexity in animals is reflected by the number of different kinds of physical parts (cells, organs, appendages). Complexity has increased over time and in particular groups by the specialization of repeated parts and by the origin of new kinds of parts. Increases in complexity of both arthropods and vertebrates have occurred in some similar ways. We have seen that the deployment of different *Hox* genes in serially repeated structures differentiates the form and function of arthropod and vertebrate structures. The success of these groups has been enabled by the flexibility of the systems governing *Hox* gene deployment such that individual structures can evolve independently of others.

The crucial insights into how this independence, and hence complexity and diversity, is achieved come from understanding the properties of genetic switches (chapter 5). Because individual genes can be and are governed by numerous independent switches, mutations in one switch can be selected for that have no effect on other switches or the function of the protein encoded by a gene. Evolutionary changes in switches are responsible for shifting the zones of *Hox* genes underlying large-scale differences in body organization in various animals (chapter 6), for finer-scale differences in the appearance of the same structure in different animals (chapters 7 and 8), and for the origin and modification of new pattern elements (chapter 8). The key to the making of "endless" forms (i.e., diversity) is in the astronomical number of possible combinations of regulatory inputs and switches. Switches integrate inputs relating to three-dimensional space, cell and tissue identity, and relative developmental time. Any of these parameters can be modified by adding, subtracting, or fine-tuning inputs into switches. Furthermore, the number of switches can expand or contract in the course of evolution. Even with a finite set of tool kit proteins that act on switches, the combinatorial power is enormous.

The realization of this power is shaped, of course, by natural selection. Not all paths are explored, not all patterns are made. Nevertheless, we delight in the 17,000 or so extant patterns of butterfly wings; the great variety of sizes, shapes, and marking of fellow mammals; the geometry of the bodies and shells of marine animals; and even the 300,000 or more species of beetles. It has been estimated that the millions of animal species now living represent perhaps just 1 percent of the billion or more forms that have evolved in the past 500 million years. We know of many diverse groups that have long since disappeared: the dinosaurs, trilobites, many weird and wonderful

Cambrian animals, and more than a dozen hominins. The combinatorial power of the genetic tool kit acting on vast arrays of genetic switches has produced this complexity and diversity.

On Novelty

EXISTING GENES AND STRUCTURES PROVIDE THE MEANS FOR INNOVATION

We have seen that insects, pterosaurs, birds, or bats did not invent "wing" genes (chapter 7), butterflies a "spot" gene (chapter 8), or humans a "bipedalism" or "speech" gene (chapter 10). Rather, innovation in all of these groups has been a matter of modifying existing structures and of teaching old genes new tricks.

The key to innovation at the genetic level is the multifunctionality of tool kit genes. The multifunctionality of tool kit genes stems from their deployment at different times and places through batteries of genetic switches. In this manner, a protein such as Distal-less can act at one time to promote limb formation, and at another to promote eyespot development. The protein made each time is identical, so the difference in function is due to its action on different switches in these different contexts.

At an anatomical level, multifunctionality and redundancy are keys to understanding the evolutionary transitions in structures. We saw this especially in arthropods, where the shifting of a function such as feeding to one of a battery of appendages freed other appendages to become specialized for walking, swimming, or other activities. In a similar fashion, the gill branches in aquatic arthropod ancestors became modified into book gills, book lungs, tubular tracheae, spinnerets, and wings.

Evo Devo has revealed the continuity among forms that was masked or about which there were uncertainties based on appearance alone. By revealing the developmental similarities among structures, Evo Devo presents a wholly new kind of evidence that is far more objective than morphology alone. These insights into the evolution of novelty strengthen aspects of Darwin's original ideas that some have found most difficult to grasp.

The history of these structures also illustrates how "endless forms" evolve through cycles of invention and expansion. New structures open up new ways of living. The insect wing led to the evolution of dragonflies and mayflies, butterflies and beetles, fleas and flies, and more. The expansion of these groups was catalyzed in turn by a cycle of innovation and expansion by making modifications to the wings or body plan—scale coloration systems in moths and butterflies, a hard covering in beetles, a sophisticated balancing hindwing in flies.

Why are existing body parts and genes the more frequent pathway to innovation? This is a matter of probability. Variation in existing structures and genes is more likely to arise than are new structures or genes, and this variation is therefore more abundant for selection to act upon. As François Jacob explained so eloquently, Nature works as a tinkerer with available materials, not as an engineer does by design. The invention of wings never occurred from scratch, but by modifying a gill branch (insects) or forelimbs (three times). Trends in evolution reflect the paths that are most available and therefore those taken most frequently.

Evo Devo has revealed that evolution can and does repeat itself at the levels of structures and patterns, as well as of individual genes. If evolution takes the most probable path, via existing structures and genes, then when confronted with similar selection pressures, different species may follow the same path to adaptation. We saw this in the evolution of feeding appendages in crustacea (chapter 6), pelvic spine reduction in sticklebacks (chapter 7), and other cases of limb reduction in vertebrates. We also saw that melanic fur or plumage patterns can arise through mutations in the very same gene in different species, and even the very same position in this gene (chapter 9).

These instances of evolution repeating itself directly address difficulties some have had in grasping the role of random mutation in the evolutionary process. Some people have found it hard to imagine how novelty and complexity arise from "a random process." The key distinction is that while the generation of genetic variation by mutation is a completely random process, the sorting of these variations as to which will persist and which will be discarded is determined by a powerful, selective nonrandom process. Of the hundreds of millions or billions of individual base pairs in an animal genome, all are equally susceptible to random copying errors or physical damage that cause mutations. But only a tiny fraction of all possible mutations can alter a mammal's coat in a viable manner, or reduce a stickleback's spines without causing catastrophic collateral damage. In large populations of animals, over eons of time, such mutations will arise simply as a matter of probability. When they do occur, positive selection upon the trait they affect will cause them to spread in populations over time.

Jacques Monod captured this interplay of randomness and selection in evolution must eloquently in the title of his landmark book, *Chance and Necessity* (a reference to the Greek philosopher Democritus who said, "Everything existing in the Universe is the fruit of chance and necessity"). Evolution is indeed a matter of chance, but in the random lottery of mutations, some numbers and combinations better meet the imperatives of ecological necessity, and they arise and are selected for repeatedly.

We also saw in rock pocket mice that the same species can use different paths to a similar solution. And, while pterosaurs, birds, and bats evolved wings out of their forelimbs, they did so in fundamentally different ways. Similar ecological demands and opportunities have selected for similar adaptations, but the developmental solutions will sometimes differ in detail.

By revealing the genetic and developmental mechanisms underlying change, Evo Devo allows us to compare and contrast the evolutionary paths of different groups. Long-standing mysteries such as Batesian mimicry in butterflies, melanism in moths, and even the evolution of finch beak size and shape now lie within our grasp. We shall soon have detailed pictures of many of the classic examples of natural selection and understand in depth how variation arises and is selected for.

On Microevolution and Macroevolution

What Is True of Species Is True of the Kingdom

The architects of the Modern Synthesis united evolutionary disciplines by asserting that the mechanisms that operated at the level of individuals in populations and species were sufficient to account for the great differences that evolve over geologic time. If, as some have proposed at various times over the past century, changes in form were due to very rare, special mutations that, for example, change a homeotic gene in just a particular way, then this extrapolation would not be justified. For a half century since the Modern Synthesis, this specter of a "hopeful monster" has lingered. The facts of Evo Devo squash it.

Evolution of homeotic genes and the traits they control has been very important, but has not occurred by different means than the sorts of mutations and variation that typically arise in populations. The preservation of *Hox* genes and other tool kit genes for more than 500 million years illustrates that the pressure to maintain these proteins has generally been as great as that upon any class of molecules. Instead, the evolutionary tinkering of switches, from those of master *Hox* genes to those of humble pigmentation enzymes, typically underlies the evolution of form. The continuity of the tool kit and the continuity of structures throughout this vast time illustrate that we need not invoke very rare or special mechanisms to explain large-scale change. The extrapolation from small-scale variation to large-scale evolution is well justified. In evolutionary parlance, Evo Devo reveals that macroevolution is the product of microevolution writ large.

Evo Devo and the Teaching of Evolution

> Don't know much about history
> Don't know much biology
> Don't know much about science books
>
> —Sam Cooke, Herb Alpert, and Lou Adler,
> "Wonderful World" (1960)

The teaching of evolution faces two challenges. The first is that it is a vast and growing subject that encompasses many disciplines. The second is that it is actively opposed, particularly in the United States, by some (but not all!) religious factions. I will address the new contribution Evo Devo can make to improving general public understanding first, the issue of opposition later.

In general, the public understanding of evolution in the United States is particularly abominable. In a survey of citizens in twenty-one countries or regions regarding general environmental and scientific knowledge, the United States placed dead last on the question of human evolution. When asked to respond to the question "In your opinion, how true is this? Human beings developed from earlier species of animals," using a four-point scale (1 represented definitely true, 2 probably true, 3 probably not true, and 4 definitely not true), the citizens of various countries responded correctly as follows:

	Nation or Area	Mean
1	East Germany	1.86
2	Japan	1.89
3	Czech Republic	2.04
4	West Germany	2.08
5	Great Britain	2.18

6	Bulgaria	2.28
7	Norway	2.43
8.5	Canada	2.45
8.5	Spain	2.45
10	Hungary	2.50
11.5	Italy	2.51
11.5	Slovenia	2.51
13	New Zealand	2.54
14	Israel	2.66
15	Netherlands	2.67
16	Ireland	2.70
17	Philippines	2.75
18	Russia	2.80
19	Northern Ireland	2.99
20	Poland	3.06
21	United States	3.22

Looking at the bright side, the United States can only move up from here.

In another survey, by the National Science Board in 1996, 52 percent of Americans polled either agreed that (32 percent) or did not know (20 percent) whether the statement "The earliest humans lived at the same time as the dinosaurs" was true.

Score that fact as two points for *The Flintstones*, zero for Darwin, Huxley, and the educational system of the world's most wealthy, powerful, and technologically driven nation.

The scandal of this ignorance is on par, I would say, with not knowing how the United States was formed, the content of its Constitution, or the roots of Western civilization. This knowledge is considered basic literacy and taught and repeated at many grade levels. So, too, is biology and earth science for which evolution must provide the basic framework. Yet the statistics are appalling.

The situation is bad enough, and reflected in other figures about science and math literacy, that the blame can probably be shared in many quarters. There are plenty of books written about and organizations studying the general problem of scientific illiteracy and its causes; I won't get into finger-pointing here. The only way up is through education. I would rather focus on what biologists and their allies at all levels of the teaching profession can do to improve matters, particularly in regard to evolution.

First, we must insist that evolution is much more than just a topic in biology—it is the foundation of the entire discipline. Biology without evolution is like physics without gravity. Just as we cannot explain the structure of the universe, the orbits of the planets and moon, or tides from mere measurement, we cannot explain human biology or Earth's biodiversity via a compendium of thousands of little facts. All general survey courses and texts must have evolution as their central unifying theme.

With respect to the scientific content to be taught, Evo Devo has much to contribute that is new, tangible, and convincing. Since the Modern Synthesis, most expositions of the evolutionary process have focused on microevolutionary mechanisms. Millions of biology students have been taught the view (from population genetics) that "evolution is change in gene frequencies." Isn't that an inspiring theme? This view forces the explanation toward mathematics and abstract descriptions of genes, and away from butterflies and zebras, or Australopithecines and Neanderthals.

The evolution of form is the main drama of life's story, both as found in the fossil record and in the diversity of living species. So, let's teach that story. Instead of "change in gene frequencies," let's try "evolution of form is change in development." This is, of course, a throwback to the Darwin-Huxley era, when embryology played a central role in the development of all evolutionary thought. There are several advantages of an embryological approach to teaching evolution.

First, it is a small leap to go from the building of complexity in one generation from egg to adult, to appreciating how increments of change in the process, assimilated over greater time periods, produce increasingly diverse forms.

Second, we now have a very firm grasp of how development is controlled. We can explain how tool kit proteins shape form, that tool kit genes are shared by all animals, and that differences in form arise from changing the way they are used. The principle of descent by modification (of development) is crystal clear.

Third, an enormous practical advantage is the visual nature of the Evo Devo perspective. The Chinese proverb I cited in chapter 4, "Hearing about something a hundred times is not as good as seeing it once," is sound educational doctrine. We learn more by combining visuals with text. Let's show students embryos, *Hox* clusters, stripes, spots, and all the glory of the making of animal form. The evolutionary concepts follow naturally.

A fourth benefit of this approach is that it brings genetics much closer to the powerful evidence of paleontology. Dinosaurs and trilobites are the poster children of evolution, and they inspire the vast majority of those who touch them. By placing these wonders of the ancient past in a continuum from the Cambrian to the present, life's history is made much more tangible. It would indeed be a wonderful world if every student had guided, repeated classroom contact with some fossils.

Let me offer a couple more general suggestions. Natural selection is often at best described as a "just so story" of adaptations: finches beaks changed due to the type of food available, moths got darker because of pollution, etc. But I do not think that the power of small increments of selection, compounded over hundreds or thousands of generations, is widely taught or understood. The commonly repeated phrase "survival of the fittest" connotes more of a gladiator contest than the subtle power of selection to act on very small differences in overall survival and fecundity. The spread of favorable mutations in populations is easily simulated and illustrated, and it underscores the time dimension of evolution.

Finally, at the university level, the evolutionary view of life should be as fundamental to a college degree as Psychology 101 or Western Civilization. But rather than asking students to memorize and regurgitate mountains of testable facts, we should emphasize study of the history of the discovery of evolution, its major characters and ideas, and the basic lines of evidence. This would do far more to inform citizens and prepare teachers than forcing students to remember the Latin names of taxa. We are stoning our children to utter boredom with little pebbles and missing the big picture. The drama of the story of evolution will recapture student interest.

There is, especially in the United States, another obstacle besides content and teaching methods to evolutionary literacy; I will address that next. But even without the active opposition, we can do better, and we have to do better.

Evo Devo and the Evolution/Creation Struggle

> If you are convinced of a matter, you must take sides or you don't deserve to succeed.
>
> —Johann Wolfgang von Goethe,
> *Propylaea* (1798)

In the short time between the first and second edition of *The Origin of Species*, Darwin inserted three more words into that famous closing paragraph, adding "by the Creator" to rewrite the phrase as "having been originally breathed by the Creator into a few forms or into one . . . " Darwin later expressed his regret for doing so in a letter to botanist J. D. Hooker: "But I have long regretted that I truckled to public opinion, and used the Pentateuchal term of creation, by which I really meant 'appeared' by some wholly unknown process."

The insertion of these words was intended to appease critics and make Darwin's evolutionary ideas more palatable. It has certainly served to fuel much speculation about Darwin's actual religious views. For some, this olive branch and Darwin's reticence in disclosing his beliefs (which are revealed to some degree only in private correspondence and unpublished notebooks) were the foundation for reconciling and accommodating evolution and religion.

Plenty of scientists and a broad spectrum of religious denominations have found such an accommodation. For example, in 1996, Pope John Paul II reiterated the Catholic position that the human body has evolved according to natural processes. Furthermore, he noted that the evidence for evolution had increased greatly, to the point where it is "more than a hypothesis." (While giving a stronger endorsement than his predecessors, the Pope is echoing a long-standing position of the Roman Catholic Church. My ordained teachers at St. Francis de Sales High School in Toledo introduced me to Darwin and evolution.) Coming from the head of the largest Christian faith in the world, with an infamously slow history of incorporating advances in science, the Pope's statements might eventually mark a turning point in evolution's long struggle for acceptance. But while some denominations have explicitly accepted the reality of biological evolution, fundamentalists who insist upon a literal reading of the Bible (referred to here as "creationists") remain firmly opposed to evolutionary science and actively promote legislation aimed at crippling the teaching of evolution in public schools.

Goethe also said, "Nothing is worse than active ignorance," and it is the agenda of these lost souls that the scientific and educational communities must thwart. I want to be very clear here in my position. I believe that the teaching of evolution and science is best served by promoting the scientific method and scientific knowledge and not by attacking religious views. The latter is a futile, counterproductive battle. However, I also believe, as many denominations have also concluded, that religion is better served by promoting and evolving its respective teachings and theologies, and not by attacking science, which is definitely a losing strategy.

Charles Harper, executive director of the John Templeton Foundation, an organization interested in the relationship between theology and science, wrote recently in the leading science journal *Nature*: "As scientific knowledge grows, religious commitments predicated on 'gaps' in scientific understanding will invariably shrink as those gaps are closed. Those Christians who are currently fighting evolutionary science will eventually need to take it seriously." Harper is right. In this time of unprecedented power in understanding embryos, genes, and genomes, and with the continual expansion of the fossil record, those gaps are fast disappearing.

One example of a mistaken faith in those gaps is that of biochemist Michael Behe, who in 1996 published *Darwin's Black Box: The Biochemical Challenge to Evolution*. Written by a credentialed scientist, Behe's book was received as a godsend by creationists. But Behe's main claim, that the living cell is an entity of irreducible complexity, is empty. Behe was counting on biology to hit a wall in reducing complex phenomena to molecular processes. He joins a long line of prognosticators whose pessimistic forecasts have been obliterated in the continuing revolution in the life sciences.

Scott Gilbert, a biologist at Swarthmore College, author of the leading college text of developmental biology, and an accomplished historian of embryology and evolutionary biology, has summarized the Behe position, and its failure: "To creationists, the synthesis of evolution and genetics cannot explain how some fish became amphibians, how some reptiles became mammals, or how some apes became human. . . . Behe named this inability to explain the creation of new taxa through genetics 'Darwin's black box.' When the box is opened, he expects evidence of the Deity to be found. However, inside Darwin's black box resides merely another type of genetics—developmental genetics."

Developmental genetics has been shedding new light on the making of complexity and the evolution of diversity for twenty years. Creationists just plain refuse to see it. How is such overt evidence ignored or dismissed? I can't pretend to understand the psychological mechanisms that allow humans to deny reality. But I do understand the desperate political and rhetorical tactics of those who, holding a losing hand, refuse to accept it. In the creationists' case, it is to assert that:

1. Evolution is *just* a theory, and there are other theories (creation or Intelligent Design) to which, out of fairness, equal treatment should be given.

or,

2. Evolution is a fraud perpetuated by scientists, or just bad science. For example, commenting on the Pope's statement, Bible-Science Association Director Ian Taylor asserted, "With the Pope's statement the Roman Catholic Church takes another step toward embracing one of the greatest deceptions ever foisted on mankind. . . . Honest scientists who know their business, such as . . . Dr. Michael Behe . . . have very forcefully pointed out, for example, that the irreducible complexity of the living cell makes chance-driven evolution absolutely impossible."

or,

3. Because scientists often disagree or are uncertain as to all mechanisms of evolution, or the relative contribution of different forces, or don't know all of the details of life's history, this uncertainty is evidence of doubt and therefore evolution is a weak theory that should not be taught.

Evolution a deception perpetuated by dishonest scientists? In their zeal, the creationists appear to have forgotten what I thought would have been one of their guiding principles, that thou shalt not bear false witness against thy neighbor.

As exasperating as the continuous battle with creationists may seem, the scientific community is now better organized and more prepared to deal with the movement. But, the battle against ignorance is not won. Rather, as Henry David Thoreau reminds us, this is a long haul:

> You can hardly convince a man of an error in a life-time, but must content yourself with the reflection that the progress of science is slow. If he is not convinced, his grandchildren may be. The geologists tell us that it took one hundred years to prove that fossils are organic and one hundred and fifty more to prove they are not to be referred to the Noachian deluge.

The struggle to advance evolutionary thought is not limited to science and scientists. Theologians such as John Haught of Georgetown University have written extensively about the need to incorporate the evolutionary perspective of science into a modern fabric of theology. Haught, who views the scientific evidence in favor of evolution as overwhelming, points out that since the text of the Bible "was composed in a prescientific age, its primary meaning cannot be unfolded in the idiom of twentieth century science" (as creationism demands). He explains:

> Many theologians have still not faced the fact that we live in a world after and not before Darwin and that an evolving cosmos looks a lot different from the world-pictures in which most religious thought was born and nurtured. If it is to survive in the intellectual climate of today, therefore, our theology requires fresh expression in evolutionary terms. When we think about God in the post-Darwinian period

we cannot have exactly the same thoughts that Augustine, Aquinas, or for the matter our grandparents and parents had. Today we need to recast all of theology in evolutionary terms.

Haught has wrestled with the significance of evolution for such theological issues as suffering, freedom, and creation. Darwin wrestled with these matters as well. Haught presents the view that creation without evolution would produce a pallid and impoverished world that would lack "all the drama, diversity, adventure, and intense beauty that evolution has produced. It might have a listless harmony to it, but it would have none of the novelty, contrast, danger, upheavals, and grandeur that evolution has in fact brought over billions of years."

This is certainly not traditional theology. But Haught's message is logical—theology must evolve, or face becoming irrelevant. We'll know it is a full-fledged revolution when fossils, genes, and embryos are discussed (positively) in Sunday school.

Endless Forms Most Endangered

The stakes in the broader adoption of an evolutionary perspective are more than philosophical. Understanding the history of our planet, both recently and in the deep past, is key to its intelligent stewardship, and to its preservation for human societies.

The evolution of *Homo sapiens* and of our culture and technology has had and is having enormous impacts on biodiversity. It is estimated that before agriculture was introduced, we numbered perhaps 10 million. Our population reached 300 million by A.D. 1 and accelerated greatly with the onset of the Industrial Revolution, reaching 1 billion around 1800. We are now at 6 billion and projected to climb to 9 billion in the next fifty years, a thousand-fold increase in just 10,000 years.

Even before the most recent boom, humans and their cultures have had dramatic effects wherever they settled. For myself, one of the most

Fig. 11.1 **The extinct thylacine.** Top, Aboriginal rock art from western Arnhem Land, Northern Territory, depicting a thylacine, a striped carnivorous marsupial long extinct from the mainland. Bottom, naturalist sketch of Tasmanian tiger (thylacine), which was still living in Tasmania in the early twentieth century. PHOTO COURTESY OF DR. CHRISTOPHER CHIPPENDALE, CAMBRIDGE UNIVERSITY

poignant examples of our impact is depicted in my most favorite place on Earth, Kakadu National Park in the Northern Territory of Australia. In addition to its spectacular diversity of living flora and fauna, Kakadu is the site of the longest continuous history of human habitation. The rock art of the Australian Aborigines is among the oldest anywhere in the world. At Ubirr, in the northern area of the park, there are many galleries with some figures drawn perhaps as long as twenty to forty thousand years ago and others made in very recent times. High up on the western face of an overhang of the main gallery is a painting of a thylacine (figure 11.1), a carnivorous marsupial also know as Van Diemen's land tiger and the Tasmanian wolf. The animal is long gone from the Northern Territory and the Australian mainland and is now completely extinct—the last thylacine from Tasmania died in a zoo in 1936. On the mainland, the thylacine was most likely the victim of dingos that came to Australia with the Aboriginal migrations. The rock art at Kakadu reminds us of what once thrived in that remarkable place, both in terms of wildlife and members of our own species.

Similar stories follow human settlement elsewhere in the world. Cave art in France portrays extinct bison and rhinoceros, piles of bones are the only remnants of the large flightless moas of New Zealand exterminated by the Maoris (figure 11.2), sketches are the only record of dodos exterminated by sailors on the island of Mauritius (figure 11.2), and the last giant ground sloths and woolly mammoths were killed off by Paleo-Indians. The quagga (figure 11.3), one of four zebra species or subspecies alive at the time of Darwin's birth, was extinct in the wild by the time of his death.

But the loss of these individual species pales in comparison to current trends of animal extinction. The large-scale destruction of habitat, the degradation of water and soil quality, the pollution of the air, and the loss of rain forests and coral reefs are wreaking global havoc on biodiversity. The butterflies and parrots of the Amazon are no longer as numerous or diverse as Bates found them, and if Darwin returned to the Galapagos Islands he would find that the very symbol

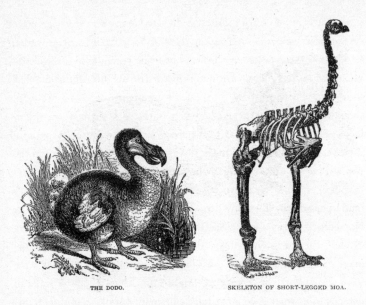

THE DODO.   SKELETON OF SHORT-LEGGED MOA.

FIG. 11.2 **The dodo and the moa.** These birds were exterminated by humans on the islands of Mauritius and New Zealand, respectively.

of the islands, the Galapagos tortoise, as well as the large ground finch and sharp-beaked ground finch, have gone extinct on some islands. Under relentless human assault, Nature's forms are not endless, nor are the most beautiful being spared.

I am not so naïve to believe that science can solve all of the world's problems, but ignorance of science, or denial of its facts, is courting doom. Recall Huxley's words when he addressed the Royal Institution at the dawn of the first revolution in biology. He asked the audience what role England would have in the grand and noble reformation of thought then under way:

Will England play this part? That depends on how you, the public, deal with science. Cherish her, venerate her, follow her methods faithfully and implicitly in their application to all branches of

FIG. 11.3 **The quagga.** This zebra species or subspecies went extinct in the late 1800s.

human thought; and the future of this people will be greater than the past. Listen to those who would silence and crush her, and I fear our children will see the glory of England vanishing like Arthur in the mist.

The question now at hand is not the glory of England, or of America, but of Nature. What a tragic irony, that the more we understand of biology, the less we have of it to learn from and to enjoy. What will be the legacy of this new century—to cherish and protect Nature, or to see butterflies and zebras and much more vanish into legend like the thylacine, moa, and dodo?

Sources and Further Reading

THE DISCOVERIES AND IDEAS discussed in this book are the fruit of many scientists' efforts. Because the presentation is intended for a broad audience, I elected not to name every individual associated with every work pertinent to the story, nor to use footnotes in the text. In this section, I provide a summary of the books and papers upon which I relied, and I offer some suggestions for further reading on particular topics of interest. In most cases, the titles of journal articles are omitted; the reference given provides enough information for interested readers to locate them.

Introduction: Butterflies, Zebras, and Embryos

The inspiration for and accounts of Darwin's, Bates's, and Wallace's travels can be found in both autobiographical and many biographical sources. I relied on Darwin's *Voyage of the Beagle* (originally published 1839; many versions and editions since) and H. W. Bates, *Naturalist on the River Amazons* (London: John Murray, 1863). For key biographical background see the Introduction by A. Shoumatoff in the Penguin

Nature Library 1988 edition of Bates's classic, pp. vii–xviii. A. Desmond and J. Moore's *Darwin: The Life of a Tormented Evolutionist* (London: Michael Joseph, 1991) is a rich source of insights into and facts of Darwin's life. The friendship between Bates and Wallace and the role it played in taking them to the Amazon is well documented and appears in nearly every capsule biography.

Many authors have commented on the aesthetic dimensions of science. Foremost among these is Robert Root-Bernstein; I strongly recommend his exceptional book *Discovering: Inventing and Solving Problems at the Frontiers of Scientific Knowledge* (Cambridge, Mass.: Harvard University Press, 1989) and his article "The Sciences and Arts Share a Common Creative Aesthetic," in *The Elusive Synthesis: Aesthetics and Science*, ed. A. Tauber, pp. 49–82 (Netherlands: Kluwer Academic Publishers, 1996). Scott Gilbert, a developmental biologist and science historian, has brought the aesthetic side of embryology to light in his article with Marion Taber, "Looking at Embryos: The Visual and Conceptual Aesthetics of Emerging Form," also in Tauber's *The Elusive Synthesis*, pp. 125–51. Among the works Gilbert and Taber single out, embryologist Paul Weiss's "Beauty and the Beast: Life and the Rule of Order," *Scientific Monthly* 81 (1955): 286–99, is an exceptional contribution.

The central role of embryology in Darwin's formulation of evolutionary ideas is evident throughout *The Origin of Species*. It also arises frequently in his correspondence—see F. Darwin, ed., *The Life and Letters of Charles Darwin*. Thomas H. Huxley, too, discusses embryos and development as important evolutionary evidence in *Evidence as to Man's Place in Nature* (1863).

The major components of the Modern Synthesis were treated in books on population genetics and evolution by Ronald A. Fisher, *The Genetical Theory of Natural Selection* (Oxford: Clarendon Press, 1930); J. B. S. Haldane, *The Causes of Evolution* (London: Longman, Green, 1932); and Theodosius Dobzhansky, *Genetics and the Origin of*

Species (New York: Columbia University Press, 1937); on systematics by Ernst Mayr, *Systematics and the Origin of Species* (New York: Columbia University Press, 1942); and on paleontology by George Gaylord Simpson, *Tempo and Mode in Evolution* (New York: Columbia University Press, 1944). Julian Huxley's *Evolution: The Modern Synthesis* (London: Allen and Unwin, 1942) integrated elements of genetics, systematics, paleontology, and botany.

Many authors have analyzed the impact and shortcomings of the Modern Synthesis, Stephen Jay Gould and Niles Eldredge, in particular. Their solo and joint works include N. Eldredge and S. J. Gould, "Punctuated Equilibria: An Alternative to Phyletic Gradualism" in *Models in Paleobiology*, ed. T. J. M. Schopf, pp. 82–115 (San Francisco: Freeman, Cooper, 1972); S. J. Gould and N. Eldredge, "Punctuated Equilibrium Comes of Age," *Nature* 366 (1993): 223–27; N. Eldredge, *Unfinished Synthesis: Biological Hierarchies and Modern Evolutionary Thought* (Oxford: Oxford University Press, 1986); and S. J. Gould, *The Structure of Evolutionary Theory* (Cambridge, Mass.: Harvard University Press, 2002). Gould's first analysis of the relationship between embryology and evolutionary processes was the landmark *Ontogeny and Phylogeny* (Cambridge, Mass.: Belknap Press, 1977).

A century before Gould's opus, Rudyard Kipling published *Just So Stories* (New York: Doubleday, 1902). One can now find many editions of these tales online.

The emergence of evolutionary developmental biology has been chronicled in a series of books for the professional and student biologist. The first was Rudy A. Raff and Thomas C. Kaufman's *Embryos, Genes, and Evolution: The Developmental-Genetic Basis of Evolutionary Change* (New York: Macmillan, 1983), which anticipated many of the fruitful questions and directions that were pursued in the decades following. More recent books include R. A. Raff, *The Shape of Life* (Chicago: University of Chicago Press, 1996); J. Gerhart and M. Kirschner, *Cells, Embryos, and Evolution* (Medford, Mass.: Blackwell

Science, 1997); E. H. Davidson, *Genomic Regulatory Systems: Development and Evolution* (San Diego: Academic Press, 2001); A. Wilkins, *The Evolution of Developmental Pathways* (Sunderland, Mass.: Sinauer Associates, 2001); and a text I cowrote with my former students Jen Grenier and Scott Weatherbee, *From DNA to Diversity: Molecular Genetics and the Evolution of Animal Design*, 2nd ed. (Medford, Mass.: Blackwell Science, 2005).

1. Animal Architecture: Modern Forms, Ancient Designs

A great introduction to the fossil fauna of Florida, including how to find them, is Mark Renz, *Fossiling in Florida: A Guide for Diggers and Divers* (Gainesville: University Press of Florida, 1999). Mark runs fossil-hunting excursions (contact Fossilx@earthlink.net) and I thank him for teaching me and my family how to search for fossils in Florida rivers and for helping us to identify what we found.

Ideas on modularity and serial repetition of structures were developed in W. Bateson, *Materials for the Study of Variation* (London: Macmillan, 1894). Williston's Law is explained in S. W. Williston, *Water Reptiles of the Past and Present* (Chicago: University of Chicago Press, 1914). More recent treatments on the importance of modularity, homology, and serial homology are G. P. Wagner, *American Zoologist* 36 (1996): 36–43, and G. P. Wagner, *Evolution* 43 (1989): 1157–71.

2. Monsters, Mutants, and Master Genes

The discovery of cyclopamine and the association of the lily *Veratrum californicum* with the induction of cyclopia are described in R. F. Keeler and W. Binns, *Teratology* 1 (1968): 5–10.

The classic experiments describing organizers in newt or frog embryos and the chick limb will be found in any modern text of developmental biology. Two such texts are Scott F. Gilbert, *Developmental*

Biology, 7th ed. (Sunderland, Mass.: Sinauer Associates, 2003), and L. Wolpert et al., *Principles of Development*, 2nd ed. (Oxford: Oxford University Press, 2002). The experiments by Spemann and his student Hilde Mangold are described in H. Spemann, *Embryonic Development and Induction* (New Haven: Yale University Press, 1938), and those of John W. Saunders and his associate M. T. Gesseling in R. Fleischmajer and R. E. Bilingham, eds., *Epithelial Mesenchymal Interactions* (Baltimore: Williams and Wilkins, 1968), pp. 78–97. The experiments describing butterfly wing eyespot organizers were first described in H. F. Nijhout, *Developmental Biology* 80 (1980): 267–74.

The term "hopeful monsters" was coined by Richard Goldschmidt in his book *The Material Basis of Evolution* (New Haven: Yale University Press, 1940). See S. J. Gould's introduction in a modern reprinting of the book (1982), pp. viii–xlii, for a discussion of the concept, as well as S. J. Gould's "Helpful Monsters" in *Hen's Teeth and Horse's Toes* (New York: W. W. Norton, 1983), pp. 187–98.

For medical descriptions of polydactyly and statistics on its occurrence, I relied upon a review by W. F. Bakker et al. in the *Electronic Journal of Hand Surgery*, November 11, 1997, accessed online, and L. G. Biesecker, *American Journal of Medical Genetics* 112 (2002): 279–83. For anecdotal accounts of individuals with polydactyly, see BaseballLibrary.com for information on Antonio Alfonseca, Wikipedia.org for historical figures, and melungeanhealth.org for a description of polydactyly in a Turkish population. For a fascinating treatment of all sorts of human developmental abnormalities, see A. M. Leroi, *Mutants: On Genetic Variety and the Human Body* (New York: Viking Press, 2003).

The literature on homeotic mutants is enormous. Short descriptions will be found in the developmental textbooks I have cited above and in Gould's "Helpful Monsters." A lengthier exploration in the full context of the development of a fly is given in Peter Lawrence, *The Making of a Fly* (Cambridge, Mass.: Blackwell Science, 1992).

3. From *E. coli* to Elephants

The origins of molecular biology, from the structure of DNA to the cracking of the genetic code, and Jacob and Monod's discoveries of the logic underlying the control of lactose metabolism in *E. coli* are detailed in Horace Freeland Judson's brilliant *The Eighth Day of Creation: The Makers of the Revolution in Biology* (New York: Simon and Schuster, 1979; reprint, with an updated preface, New York: Cold Spring Harbor Laboratory Press, 1996). This is one of the best written, most thoroughly researched books in the entire genre of science writing.

Explanations of how genetic information is encoded and decoded will be found in most college-level biology textbooks, and an online search for the keywords "DNA, RNA, and proteins" will lead to numerous short illustrated synopses. The control of beta-galactosidase production is detailed in most textbooks on genetics and molecular biology, and in a compendium of papers in J. H. Miller and W. S. Reznikoff, eds., *The Operon* (Cold Spring Harbor, N.Y.: Cold Spring Harbor Laboratory Press, 1978).

The books by Jacques Monod and François Jacob to which I refer are J. Monod, *Chance and Necessity* (New York: Alfred A. Knopf, 1971); F. Jacob, *The Logic of Life* (New York: Pantheon, 1974); and F. Jacob, *The Statue Within: An Autobiography* (New York: Basic Books, 1988). François Jacob has also more recently written a book on advances in genetics and developmental biology, including the homeobox story, entitled *Of Flies, Mice, and Men: On the Revolution in Modern Biology by One of the Scientists Who Helped Make It* (Cambridge, Mass.: Harvard University Press, 1998).

Key papers on the genetics of the Antennapedia and Bithorax Complexes were E. Lewis, *Nature* 276 (1978): 565–70, and B. Wakimoto and T. Kaufman, *Developmental Biology* 81 (1981): 51–64. The homeobox was discovered in parallel by individuals working in two laboratories, one headed by Thom Kaufman at Indiana University, the other headed by Walter Gehring at the University of Basel.

Accounts of the discovery can be found in Peter Lawrence, *The Making of the Fly* (Medford, Mass.: Blackwell Science, 1992), in W. Gehring, *Master Control Genes in Development and Evolution: The Homeobox Story* (New Haven: Yale University Press, 1999), and in W. McGinnis, *Genetics* 137 (1994): 607–11. The primary references are M. P. Scott and A. J. Weiner, *Proceedings of the National Academy of Science, USA* 81 (1984): 4115–19, and W. McGinnis et al., *Nature* 308 (1984): 428–33. The report of the similarity between the homeodomain and well-known bacterial and yeast regulatory proteins is A. S. Laughon and M. P. Scott, *Nature* 310 (1984): 25–31. The discovery of homeobox genes in other animals was reported in W. McGinnis et al., *Cell* 37 (1984): 403–8. Jonathan Slack's article comparing the homeobox to the Rosetta stone appeared in *Nature* 310 (1984): 364–65. A commentary by Stephen Jay Gould on the significance of the homeobox appeared in *Natural History* 94 (1985): 12–23.

The discovery of the features of *Hox* gene organization in clusters and their expression along the body axes in vertebrates was reported by D. Duboule and P. Dollé, *EMBO Journal* 8 (1989): 1497–1505, and A. Graham, N. Papalapov, and R. Krumlauf, *Cell* 57 (1989): 367–78.

The homology of the *Drosophila eyeless* gene to the *Small eye* and *Aniridia* genes of mice and humans, respectively, was reported by R. Quiring et al., *Science* 265 (1994): 785–89, and the ability of the *eyeless* and *Small eye* gene products to induce eye tissue at additional sites in the fly was reported by G. Halder, P. Callaerts, and W. Gehring, *Science* 267 (1994): 1788–92. A commentary on this work was written by S. J. Gould, *Natural History* 103 (1994): 12–20. Richard Dawkins has written a terrific explanation of the evolution of eyes, "The Forty-Fold Path to Enlightenment," in his *Climbing Mount Improbable* (New York: W. W. Norton, 1996), pp. 138–97.

The use of the *Distal-less* gene and its homologs in the formation of many kinds of appendages was reported by G. Panganiban et al., *Proceedings of the National Academy of Science, USA* 94 (1997): 5162–66. A discussion of the significance of the role of *tinman* and

NK2 homeobox genes in the building of fly and vertebrate hearts is offered by R. Bodmer and T. V. Venkatregh, *Developmental Genetics* 22 (1998):181–86.

Enrst Mayr's view of evolutionary distance can be found in his *Animal Species and Evolution* (Cambridge, Mass.: Harvard University Press, 1963), p. 609. Stephen Jay Gould's comments in *The Structure of Evolutionary Theory* (Cambridge, Mass.: Harvard University Press, 2002) are found on p. 1065.

The first report of Nüsslein-Volhard and Wieschaus's pioneering search for the genes that sculpt the pattern of the fruit fly embryo appeared in *Nature* 287 (1980): 795–801. It was many years later when the *Drosophila hedgehog* gene was isolated and, shortly therafter, its vertebrate homologs. The report of the ability of the Sonic hedgehog protein to mimic the activity of the ZPA in the chick limb was R. Riddle et al., *Cell* 75 (1993): 1401–16. The association of mutations in *Sonic hedgehog* with polydactyly in humans was reported by L. Lettice et al., *Proceedings of the National Academy of Science, USA* 99 (2002): 7548–53.

The induction of cyclopia by mutations in the *Sonic hedgehog* gene was reported by C. Chiang et al., *Nature* 383 (1996): 407–13. This observation, coupled with the discovery that some cancers are associated with mutations in other genes in the pathway, led to the testing of cyclopamine as a potential chemotherapy; see J. Taipale et al., *Nature* (2000): 1005–9, and commentary by A. E. Bale, *Nature* 406 (2000): 944–45.

4. Making Babies: 25,000 Genes, Some Assembly Required

The title wordplay was suggested by an anecdote in Scott Gilbert and Marion Taber's "Looking at Embryos: The Visual and Conceptual Aesthetics of Emerging Forms" in *The Elusive Synthesis: Science and Aesthetics,* ed. A. Tauberg, pp. 125–51 (Netherlands: Kluwar Academic Publishers, 1996). They mentioned that in 1992 the *Encyclopedia of the Mouse Genome* carried the banner "The Complete Mouse" with "some assembly required" in parentheses. Gilbert and Taber are also the source of the Paul Weiss anecdote about getting the chicken back.

The analogy of embryology to mapmaking is explained in Stephen S. Hall, *Mapping the Next Millennium: The Discovery of New Geographies* (New York: Random House, 1992), pp. 193–214. Hall makes a compelling case throughout the book of the central role of mapmaking in the sciences; he is right on point with regard to genetics, embryology, and genomics.

The processes of embryonic development are also explained and illustrated in two books written for general audiences by great developmental biologists. Lewis Wolpert's *Triumph of the Embryo* (New York: Oxford University Press, 1991) is a concise and very clear outline of the key steps in making embryos and structures. Enrico Coen's *Art of the Genes: How Organisms Copy Themselves* (Oxford: Oxford University Press, 1998) has a unique perspective that intertwines embryology and art in illuminating how patterns are encoded and revealed.

Fate mapping is described in all developmental biology textbooks, including those I cited earlier for chapter 2. An excellent recent review of goals, strategies, and new methodologies in fate mapping is J. D. W. Clarke and C. Tickle, *Nature Cell Biology* 1 (1999): E103–9. Figures 4.1 and 4.2 are simplified from several sources including the above references and the fate maps of Volker Hartenstein published as a supplementary atlas in M. Bate and A. Martinez-Arias, eds., *The Development of Drosophila melanogaster* (Cold Spring Harbor, N.Y.: Cold Spring Harbor Laboratory Press, 1993).

The descriptions of tool kit gene expression are drawn from work in my laboratory, many primary literature reports, information provided by colleagues who contributed images, as well as textbook sources. Again, the developmental biology textbooks cited will contain most of this information in much greater detail. Peter Lawrence, *The Making of the Fly* (Cambridge, Mass.: Blackwell Science, 1992), contains much detail about genes expressed in the fly embryo, and Sean B. Carroll, Jen

Grenier, and Scott Weatherbee, *From DNA to Diversity: Molecular Genetics and the Evolution of Animal Design*, 2nd ed. (Medford, Mass.: Blackwell Science, 2005), also describes the steps involved in building flies and vertebrates. Some reviews that cover particular topics in depth include: for early vertebrate embryos, E. M. De Robertis et al., *Nature Reviews Genetics* 1 (2000): 171–81; the making of segments in vertebrates, O. Pourquie, *Science* 301 (2003): 328–30; the building of the vertebrate limb, F. Moriani and G. R. Martin, *Nature* 423 (2003): 319–25; the making of the hindbrain, C. B. Moens and V. E. Prince, *Developmental Dynamics* 224 (2002): 1–17.

The description of lateral inhibition is generalized from numerous examples, such as the positioning of bristles and feather buds. This concept is discussed in detail and numerous examples reviewed in H. Meinhardt and A. Gierer, *BioEssays* 22 (2002): 753–60. There are some excellent tutorials and animations on these authors' Web site concerning the generation of periodic and spacing patterns: www.eb.tuebingen.mpg.de/dept4/meinhardt/home.html.

François Jacob's quotation of Jean Perrin appears in his essay "Evolution and Tinkering," *Science* 196 (1977): 1161–66. Jean Perrin was a Nobel laureate in Physics (1926) who was cited for his work on colloids and Brownian motion. He wrote a very popular book, *Les Atomes* (1913), from which the quotation is taken.

5. The Dark Matter of the Genome:
Operating Instructions for the Tool Kit

I first encountered "dark matter" in Brian Greene's *The Elegant Universe* (New York: W. W. Norton, 1999), a very engaging book about the structure of the universe from the very smallest to the largest scale, and in Martin Rees's excellent *Just Six Numbers: The Deep Forces That Shape the Universe* (New York: Basic Books, 2001). For additional articles, see Dennis Overbye, "From Light to Darkness:

Astronomy's New Universe," *The New York Times*, April 10, 2001, and Vera Rubin, *Scientific American Presents* 9, no. 1 (1998), 106–10.

There are several texts that focus on the properties of genetic switches in detail. Mark Ptashne's classic *A Genetic Switch*, 2nd ed. (Cambridge, Mass.: Blackwell Science, 1992), is a short, very well illustrated, step-by-step tutorial on genetic switches, primarily in bacteriophage, but with some examples from more complex organisms. Eric H. Davidson, *Genomic Regulatory Systems: Development and Evolution* (San Diego: Academic Press, 2001) is an advanced text that explains the logic and operations of the more complicated switches of animal genes.

Estimates of the amount of "junk" DNA in our genome and the fraction that is regulatory are based upon studies of the human genome sequence, and comparisons with other species, particularly the mouse, as described by the Mouse Genome Sequencing Consortium, *Nature* 420 (2002): 520–62.

Key references for the operation of genetic switches in positioning stripes and clusters of cells are: D. Stanojevic, S. Small, and M. Levine, *Science* 254 (1991): 1385–87; S. Small, A. Blair, and M. Levine, *EMBO Journal* 11 (1992): 4047–57; G. Vachon et al., *Cell* 71 (1992): 437–50; J. Jiang and M. Levine, *Cell* 72 (1993): 741–52; S. Gray, P. Szymanski, and M. Levine, *Genes and Development* 8 (1994): 1829–38; S. Gray and M. Levine, *Genes and Development* 10 (1996): 700–710; P. Szymanski and M. Levine, *EMBO Journal* 14 (1995): 2229–38; and J. Cowden and M. Levine, *Developmental Biology* 262 (2003): 335–49. The signature sequences bound by particular tool kit proteins are derived from these references and S. Jun et al., *Proceedings of the National Academy of Sciences, USA* 95 (1998): 13720–725; S. Knirr and M. Frasch, *Developmental Biology* 238 (2001): 13–26; and S. C. Ekker et al., *EMBO Journal* 13 (1994): 3551–60.

Examples of Turing-like models of pattern formation are discussed in S. Kauffman's *The Origins of Order* (Oxford: Oxford University Press, 1993) and P. Ball, *The Self-Made Tapestry: Pattern Formation in Nature* (Oxford: Oxford University Press, 1999). It is interesting to compare the analyses of fly development in the two books. Kauffman's treatment is longer and more complex and did not incorporate the discovery of switches for making individual stripes (made a couple of years before publication). Ball's treatment is greatly simplified and clarified by the description of how switches transform fuzzy patterns into sharper patterns. Still, the central importance of genetic switches to pattern formation has not yet fully penetrated the computational modeling world; for example, see S. Wolfram, *A New Kind of Science* (Champaign, Ill.: Wolfram Media, 2002). The continuing mistake is being seduced into believing that simple rules that can generate patterns on a computer screen are the rules that generate patterns in biology.

For information on the genetic switches of the *BMP5* gene I relied on personal communications with Dr. David Kingsley of Stanford University and on R. J. Di Leone et al., *Proceedings of the National Academy of Sciences, USA* 97 (2000): 1612–17. The logic of how Hox proteins and other tool kit proteins differentiate modules in animal bodies is summarized in S. D. Weatherbee and S. B. Carroll, *Cell* 97 (1999): 283–86.

6. The Big Bang of Animal Evolution

There are several excellent books for the general reader that deal entirely or in part with the Cambrian Explosion. S. J. Gould, *Wonderful Life: The Burgess Shale and the Nature of History* (New York: W. W. Norton, 1989) was the first to bring the phenomenon of the Cambrian to a wide audience. Simon Conway Morris, *The Crucible of Creation: The Burgess Shale and the Rise of Animals* (New York: Oxford University Press, 1998) tells the story from the viewpoint of one of the leading paleontologists working on the fossils, and is more up-to-date with respect to interpretation and the inclusion of insights from other Cambrian sites. Andrew H. Knoll, *Life on a Young Planet: The First Three Billion Years of Evolution on Earth* (Princeton: Princeton University Press, 2003) covers the entire known history of life up to and including the Cambrian—it is a terrific synthesis of geology, geochemistry, and paleontology. Derek E. G. Briggs, Douglas H. Erwin, and Frederick J. Collier, *The Fossils of the Burgess Shale* (Washington, D.C.: Smithsonian Institution Press, 1994) is a handsome catalog describing Burgess fossils.

The first description of Urbilateria was E. M. De Robertis and Y. Sasai, *Nature* 380 (1996): 37–40. Additional articles pertaining to Urbilateria are: De Robertis, *Nature* 387 (1997): 25–36; C. B. Kimmel, *Trends in Genetics* 12 (1996): 329–31; N. Shubin, C. Tabin, and S. Carroll, *Nature* 388 (1997): 639–48; D. Arendt and J. Wittbrodt, *Philosophical Transactions of the Royal Society of London* B 350 (2001): 1545–63; D. Arendt, U. Technau, and J. Wittbrodt, *Nature* 409 (2001): 81–85; A. H. Knoll and S. B. Carroll, *Science* 284 (1999): 2129–37; D. H. Erwin and E. H. Davidson, *Development* 129 (2002): 3021–32; and A. Peel and M. Akam, *Current Biology* 18 (2003): R708–10.

The quote from Darwin on mankind's genealogy comes from his letter to Charles Lyell of January 10, 1860, published in *The Life and Letters of Charles Darwin*, ed. F. Darwin, vol. 2 (London: John Murray, 1887).

For background on the evolution of lobopodians, I relied upon G. E. Budd, *Lethaia* 29 (1996): 1–14, and personal communications with Dr. Graham Budd of the University of Uppsala, Sweden.

Lewis's "new genes" hypothesis appears in E. B. Lewis, *Nature* 276 (1978): 565–70. The description of the *Hox* genes of Onychophorans is J. K. Grenier et al., *Current Biology* 7(1997): 547–53. The literature on the shifting of *Hox* zones in arthropods is substantial and growing; key references are M. Averof and M. Akam, *Nature* 376 (1995): 420–23; S. B. Carroll, *Nature* 376 (1995): 479–85. M. Averof and N. H. Patel, *Nature* 388 (1997): 682–87; C. L. Hughes and T. C. Kaufman, *Development* 129 (2002): 1225–38; and N. C. Hughes, *BioEssays* 28 (2003): 386–95.

The detailed description of *Haikouichthys* is in D. G. Shu et al., *Nature* 421 (2003): 526–29. The analysis of cephalochordate *Hox* genes is in J. Garcia-Fernandez and P. W. Holland, *Nature* 370 (1994): 563–66;

of lamprey and hagfish *Hox* genes in H. Ecriva et al., *Molecular and Biological Evolution* 19 (2002): 1440–50, and C. Fried, S. J. Prohaska, and P. F. Stadler, *Journal of Experimental Zoology Part B Molecular and Developmental Evolution* 299 (2003): 18–25; and of sharks in C.-B. Kim et al., *Proceedings of the National Academy of Sciences, USA* 97 (2000): 1055–60. The innovations of vertebrates are discussed in S. M. Shimeld and P. W. Holland, *Proceedings of the National Academy of Sciences, USA* 97 (2000): 4449–52, and G. P. Wagner, C. Amemiya, and F. Ruddle, *Proceedings of the National Academy of Sciences, USA* 100 (2003): 14603–606. The expression of *Hox* genes in different vertebrates is detailed in A. C. Burke et al., *Development* 121(1995): 333–46, and M. J. Cohn and C. Tickle, *Nature* 399 (1999): 474–79. The evolution of a vertebrate *Hox* gene switch is reported in H.-G. Belting, C. Shashikant, and F. H. Ruddle, *Proceedings of the National Academy of Sciences, USA* 95 (1998): 2355–60.

For a discussion of the role of ecology in the Cambrian, see Knoll's *Life on a Young Planet*.

7. Little Bangs: Wings and Other Revolutionary Inventions

The histories of the knife and fork and paper clips are described in A. B. Duthie, *Journal of Memetics—Evolutionary Models of Information Transmission* 8 (2003), available at http://jom-emit.cfpm.org/2004/vol8/duthie_ab.html, and H. Petroski, *The Evolution of Useful Things* (New York: Vintage Books, 1992). Darwin's discussion of the importance of multifunctionality and redundancy is in chapter 6, "Difficulties of the Theory," in *The Origin of Species*.

The structure and importance of biramous limbs are discussed at length in S. J. Gould, *Wonderful Life: The Burgess Shale and the Nature of History* (New York: W. W. Norton, 1989), and scenarios for their origin are discussed in G. E. Budd, *Lethaia* 29 (1996): 1–14, and N. Shubin, C. Tabin, and S. Carroll, *Nature* 388 (1997): 639–48. The expression of the *Distal-less* limb-building gene in arthropod and

Onychophoran limbs is reported in G. Panganiban et al., *Science* 270 (1995): 1363–66, and Panganiban et al., *Proceedings of the National Academy of Sciences, USA* 94 (1997): 5162–66.

The evidence for the evolution of the insect wing from the gill branch of an aquatic ancestor is M. Averof and S. M. Cohen, *Nature* 385 (1997): 627–30. The scenario for the evolution of insect wing number is derived from S. B. Carroll, S. D. Weatherbee, and J. A. Langeland, *Nature* 375 (1995): 58–61, and based partly on fossil evidence in J. Kukalova-Peck, *Journal of Morphology* 156 (1978): 53–126.

The evidence for the evolution of spider spinnerets, book lungs, and tubular tracheae from an ancestral gill branch is W. G. M. Damen, T. Saridaki, and M. Averof, *Current Biology* 12 (2002): 1711–16. The different *Hox* zones in spiders are reported in W. G. M. Damen et al., *Proceedings of the National Academy of Sciences, USA* 95 (1998): 10665–670, and A. Abzhanov, A. Popadic, and T. C. Kaurman, *Evolution and Development* 1 (1999): 77–89. The evolution of the insect hindwing under the control of the Ultrabithorax protein is based upon S. D. Weatherbee et al., *Current Biology* 11 (1999): 109-15.

For a detailed treatment of the evolution of the vertebrate limb, both in adapting to land and back to water, see Carl Zimmer, *At the Water's Edge: Macroevolution and the Transformation of Life* (New York: Free Press, 1998). Descriptions of *Sauripteris, Acanthostega, Tulerpeton*, and other fossils are in E. B. Daeschler and N. Shubin, *Nature* 391 (1998): 133, and M. I. Coates, J. E. Jeffrey, and M. Rut, *Evolution and Development* 4 (2002): 390–401. The evolution of the autopod with respect to *Hox* genes is described in P. Sardino, F. van der Hoeven, and D. Duboule, *Nature* 375 (1995): 678–81; N. Shubin, C. Tabin, and S. Carroll, *Nature* 388 (1997): 639–48; and M. Kmita et al., *Nature* 420 (2002): 145–50.

The evolution of the different forms of vertebrate wings is the focus of Pat Shipman's *Taking Wing: Archeopteryx and the Evolution of Bird Flight* (New York: Simon and Schuster, 1998). The developmental basis of limblessness in snakes is reported by M. J. Cohn and C. Tickle, *Nature* 399 (1999): 474–79. The evolution of spine reduction in threespine stickleback fish is described in M. D. Shapiro et al., *Nature* 428 (2004): 717–23, and the exceptional high-resolution fossil record of sticklebacks is described in M. A. Bell, J. V. Baumgartner, and E. C. Olsen, *Paleobiology* 11 (1985): 258–71.

8. How the Butterfly Got Its Spots

Although the quote at the beginning of the chapter is often cited, Monod never said it—which bears some explanation. In *Le Hasard et la Nécessité* (Paris: Editions du Seuil, 1970), Monod wrote, *"Hasard capté, conserve, reproduit per la machinerie de l'invariance et ainsi converti en ordre, régle, nécessité"* (p. 128). His English-language translator, Austryn Wainhouse, chose to translate *"hasard capté"* as "randomness caught on the wing," but more literally it would be "chance [or, randomness] captured." Stuart Kauffman, in *At Home in the Universe* (New York: Oxford University Press, 1995), first quotes Monod as saying "chance caught on the wing" (p. 71) and later extends the quote to "Evolution is chance caught on the wing" (p. 97). A wonderful phrase, well worth quoting, but neither Monod nor his translator ever wrote it.

The statistics on Bates's collections are in H. W. Bates, *Naturalist on the River Amazons* (London: John Murray, 1863). The letter quoted from Bates to Darwin was received on March 28, 1861. Darwin's enthusiastic comment on Bates's paper on mimicry was written on November 20, 1862, and appears in *The Life and Letters of Charles Darwin*, ed. F. Darwin, vol. 2 (London: John Murray, 1887). Darwin's appreciation of Bates's book was published in the *Natural History Review* 3 (1863). All of the quoted passages about butterflies are from *Naturalist on the River Amazons*. For more on Nabokov, see K. Johnson and S. Coates, *Nabokov's Blues: The Scientific Odyssey of a Literacy Genius* (Cambridge, Mass.: Zoland, 1999).

The most comprehensive analysis of butterfly wing patterns is H. Frederik Nijhout, *The Development and Evolution of Butterfly Wing Patterns* (Washington, D.C.: Smithsonian Institution Press, 1991), which explains much of the background I cover on the structure and diversity of wing patterns. The tool kit gene associated with the development of scales is reported in R. Galant et al., *Current Biology* 8 (1998): 807–13.

The discovery of *Distal-less* gene expression in spots in developing wings is S. B. Carroll et al., *Science* 265 (1994): 109–14; see also S. B. Carroll, *Natural History*, February 1997, pp. 28–37. The comparison of Distal-less expression in different species is illustrated in P. M. Brakefield et al., *Nature* 384 (1996): 236–42. The tool kit proteins that mark the outer rings of eyespots are reported in C. R. Brunetti et al., *Current Biology* 11 (2001): 1578–85.

The role of reduced eyespots for hiding on dead leaf litter is examined in A. Lytinen et al., *Proceedings of the Royal Society of London* B 271 (2004): 279–83. The expression of *Distal-less* in butterflies reared at different temperatures is reported in P. M. Brakefield et al., *op. cit.* The *Spotty* mutant is described in P. M. Brakefield and V. French, *Acta Biotheoretica* 41 (1993): 447–68. The use of artificial selection to evolve lines of butterflies with different eyespot sizes is described in A. F. Monteiro, P. M. Brakefield, and V. French, *Evolution* 48 (1994): 1147–57. A general overview of recent studies of butterfly wing-pattern evolution is P. Beldade and P. M. Brakefield, *Nature Reviews Genetics* 3 (2002): 442–52.

Mimicry in the tiger swallowtail is described in J. M. Scriber, R. H. Hagen, and R. C. Lederhouse, *Evolution* 50 (1996): 222–36. There is a large literature on mimicry in *Heliconius* butterflies; see J. Mallet and M. Joron, *Annual Rev. Ecol. Syst.* 30 (1999): 201–33.

9. Paint It Black

For more work by Hugh B. Cott, see *The Royal Engineers Journal* 52 (1938): 501–17, and *Looking at Animals: A Zoologist in Africa* (New York: Charles Scribner Sons, 1975).

For a broad discussion of melanism see M. Majerus, *Melanism: Evolution in Action* (Oxford: Oxford University Press, 1988). Recent scrutiny of the biology of industrial melanism in peppered moths is provided by B. N. Grant, *Evolution* 53 (1999): 980–84, and J. Mallet, *Genetics Society News* 50 (2003): 34–38; the latter responds to J. Hopper, *Of Moths and Men: Intrigue, Tragedy, and the Peppered Moth* (New York: Fourth Estate, 2002).

An excellent review of melanism in mammals is M. E. N. Majerus and N. I. Mundy, *Trends in Genetics* 19 (2003): 585–88. The primary references are: for the jaguar and jaguarundi, E. Eizirik et al., *Current Biology* 13 (2003): 448–53; for the bananaquit, E. Theron et al., *Current Biology* 11 (2001): 550–57; for the rock pocket mouse, M. Nachman et al., *Proceedings of the National Academy of Science, USA* 100 (2003): 5268–73; and for the Kermode bear, K. Ritland et al., *Current Biology* 11 (2001): 1468–72. Field studies of the rock pocket mice of the deserts of the American southwest are L. Dice and P. M. Blossom, *Studies of Mammalian Ecology in Southwestern North American with Special Attention for the Colors of Desert Mammals* (Washington, D.C.: Carnegie Institution of Washington, 1937), pub. no. 485, and L. R. Dice, *Contributions from the Laboratory of Vertebrate Biology* (University of Michigan) 34 (1947): 1–20.

Gould's essays on zebras are in *Hen's Teeth and Horse's Toes* (New York: W. W. Norton, 1983), pp. 355–65 and 366–75. J. L. Bard's analysis is in *Journal of Zoology (London)* 183 (1977): 527–39.

The general formulas for the time required for advantageous mutants to spread in a population, or the probability of disadvantageous mutations being lost from a population, are in most any population genetics text; see, for example, W.-H. Li, *Molecular Evolution* (Sunderland, Mass.: Sinauer Associates, 1997).

10. A Beautiful Mind: The Making of *Homo sapiens*

Darwin's reactions to observing orangutans are described in A. Desmond and J. Moore, *Darwin: The Life of a Tormented Evolutionist* (New York: Warner, 1997). Queen Victoria's diary entry on Jenny is quoted in R. A. Keynes, *Annie's Box* (London: Fourth Estate, 2001). The quote from Erich Fromm appeared in his *Man for Himself* (New York: Rinehart, 1947).

A broad overview of the physical and genetic history of human evolution is found in J. Klein and N. Takahata, *Where Do We Come From? The Molecular Evidence for Human Descent* (Berlin: Springer-Verlag, 2002). Some of the topics addressed here are examined in S. B. Carroll, *Nature* 422 (2003): 849–57.

The story of the discovery of the first Neanderthal is told in R. McKie, *Dawn of Man: The Story of Human Evolution* (London: Dorling Kindersley, 2000), and its first meaningful interpretation was detailed in T. H. Huxley, *Evidence as to Man's Place in Nature* (1863). *The Athenaeum*'s review of Huxley's book appeared on February 28, 1863. The oldest *H. sapiens* specimen is described in T. D. White et al., *Nature* 423 (2003): 742–47.

The data represented in figures 10.3 and 10.5 are compiled from many sources. I have received guidance from paleontologists with differing views on the number and identity of distinct hominin species. I chose a more conservative rather than all-inclusive picture here. For differing views, see B. Wood, *Nature* 418 (2002): 133–35, and T. White, *Science* 299 (2003): 1994–96.

For more information on the footprints at Laetoli, see N. Agnew, *Scientific American* 279 (1998): 51–54. Data on fossil brain sizes are derived from R. B. Ruff, E. Trinkhaus, and T. Holliday, *Nature* 387 (1997): 173–76; G. Conroy et al., *American Journal of Physical Anthropology* 13 (2000): 111–18; P. Brunet et al., *Nature* 418 (2002): 145–51; and B. Wood, *Science* 284 (1999): 65–71. J. M. Allman, *Evolving Brains* (New York: Scientific American Library, 1999) was a source of much information on primate and human brain structure and evolution, behavior, and climatic change. The mosaic pattern of brain evolution is described by R. A. Barton and P. Harvey, *Nature* 408 (2000): 1055–58; W. de Winter and C. E. Oxnard, *Nature* 409 (2001): 710–14a; and D. A. Clark, P. P. Mitra, and S. S. H. Wang, *Nature* 411 (2001): 189–93.

The first study of Neanderthal DNA was M. Krings et al., *Cell* 90 (1997): 19–30; see also D. Serre et al., *Public Library of Science/Biology* 2 (2004): 0313–0317.

The quote from Emerson Pugh is from his *The Biological Origin of Human Values* (New York: Basic Books, 1977).

Evidence for anatomical asymmetries in great ape brains is discussed in C. Cantalupo and W. D. Hopkins, *Nature* 414 (2001): 505, but is strongly countered by C. C. Sherwood et al., *The Anatomical Record Part A* 271 (2003): 276–85. Studies of patients with situs invertus are D. Kennedy et al., *Neurology* 53 (1999): 1260–65, and S. Tanaka et al., *Neuropsychologia* 37 (1999): 869–74.

The arithmetic of human DNA sequence evolution is derived from the full human genome sequence and comparable data available for chimpanzees—see, for instance, S. B. Carroll, *Nature* 422 (2003): 849–57. The comparison with the mouse is based upon the Mouse Genome Squence Consortium, *Nature* 420 (2002): 520–62, and an update presented by Dr. Eric Lander, Breckinridge, Colorado, January 2004.

The classic reference on human-chimpanzee differences is M.-C. King and A. C. Wilson, *Science* 188 (1975): 107–16. Additional early views were E. Zuckerkandl and L. Pauling in *Evolving Genes and Proteins*, ed. V. Bryson and J. H. Vogel, pp. 97–166 (New York: Academic Press, 1965), and R. J. Britten and E. H. Davidson, *Quarterly Review of Biology* 46 (1971): 111–38.

The association of a myosin gene mutation with reduction in human jaw musculature is reported by H. Stedman et al., *Nature* 428 (2004): 415–18.

The discovery of the *FOXP2* gene associated with a speech and language disorder is in C. S. L. Lai et al., *Nature* 413 (2001): 519–23; imaging of patients with the disorder is described by F. Liégeois et al., *Nature Neuroscience* 6 (2003): 1230–36; the molecular evolution of the *FOXP2* sequence is reported in W. Enard et al. *Nature* (2002) 418: 869-872; *FOXP2* expression in the human brain is reported in C. S. Lai et al., *Brain* 126 (2003): 2455–62 ; *FOXP2* expression in rats and mice brains is described in K. Takahashi et al., *Journal of Neuroscience Research* 73 (2003): 61–72, and R. J. Ferland et al., *Journal of Comprehensive Neurology* 460 (2003): 266–79.

For more on genes, experience, and human behavior, see M. Ridley, *Nature via Nurture: Genes, Experience, and What Makes Us Human* (New York: HarperCollins, 2003).

11. Endless Forms Most Beautiful

The earlier versions of passages in *The Origin of Species* are found in *The Foundations of the Origin of Species: Two Essays Written in 1842 and 1844 by Charles Darwin*, ed. Francis Darwin (Cambridge: Cambridge University Press, 1909).

For additional perspective on the tendency of evolution to repeat itself at various levels, see Simon Conway Morris's *Life's Solution: Inevitable Humans in a Lonely Universe* (Cambridge: Cambridge University Press, 2003).

Data on the public understanding of evolution are from G. Bishop, *The Public Perspective* 9 (1998): 39–44. Further information on the state of evolution education can be found on the Web site of the National Center for Science Education (www.natcenscied.org).

Detailed information on the various changes made between editions of *The Origin of Species* can be found in Morse Peckham, ed., *The Origin of Species by Charles Darwin: A Variorum Text* (Philadelphia: University of Pennsylvania Press, 1959). Pope John Paul II's statement on evolution and various individual reactions to it are discussed in E. C. Scott, *The Quarterly Review of Biology* 72 (1997): 401–6. Charles Harper's comments appear in *Nature* 411 (2001): 239–40. Scott Gilbert's views on teaching evolution through developmental genetics and his critique of M. Behe's *Darwin's Black Box: The Biochemical Challenge to Evolution* (New York: Free Press, 1996) appear in *Nature Reviews Genetics* 4 (2003): 735–41.

Thoreau talks of the long haul in his 1849 *A Week on the Concord and*

Merrimack Rivers. The views of John F. Haught are from *Science and Religion: From Conflict to Conversation* (New York: Paulist Press, 1995). Excerpts of his writings can be found at www.aaas.org/spp/dser/ evolution/perspectives/haught.shtml.

The statistics on population growth and its history are taken from the Population Reference Bureau (www.prb.org).

The story of the thylacine is told in D. Owen, *Thylacine: The Tragic Tale of the Tasmanian Tiger* (Crows Nest, NSW: Allen and Unwin, 2003). Further information on species extinction is found in E. O. Wilson and F. M. Peter, eds., *Biodiversity* (Washington, D.C.: National Academy Press, 1988), and E. O. Wilson, *The Diversity of Life* (New York: Penguin, 1992).

Huxley's address to the Royal Institution in February 1860 is quoted in A. Desmond and J. Moore, *Darwin: The Life of a Tormented Evolutionist* (New York: Warner, 1991), p. 489.

Acknowledgments

AS I HOPE IS TRUE for any author, this book has been a labor of love. But I have certainly been luckier than most because my labors were made possible and lightened enormously by the help of my wife, Jamie Carroll. Her critical taste and encouragement nourished the book's birth, her hard work and artistic talent fed its growing body, her patience endured countless questions of "Honey, what do you think of this quote/paragraph/section/chapter/title/picture/etc.?," and her honest answers spared readers much confusion and pain. No one could ever hope for a more generous partner, a warmer home in which to create, or a better sense of humor for getting through the inevitable twists and turns.

The research for and the making of the book were aided in many ways by my wonderful family, who have trudged happily through various jungles, swamps, muddy rivers, and innumerable museums for the love of natural history. My sons, Will and Patrick, helped to find fossils in the field and key animals in museums and my stepson, Josh Klaiss, created several important graphics.

I thank my sister, Nancy, with whom I have studied and discussed the lives of Darwin, Huxley, Lyell, and their contemporaries for almost a decade, my brother Peter for always pushing for the big picture and our many discussions on human evolution, and my brother Jim for his great encouragement.

I also thank my parents, Joan Carroll and the late J. Robert Carroll, for encouraging each of their children to pursue whatever interested us, even when that meant keeping snakes in the house.

The artwork here was a big undertaking. Original drawings and graphics were created by Jamie, Josh, and Leanne Olds. Leanne also composed or redrew most of the figures that originated from other sources. Steve Paddock, a longtime member of my research group, compiled and arranged the color artwork. I am grateful for the care everyone put into each image and I am thrilled at the results.

Much of the artwork was contributed by colleagues around the world and is the fruit of their field and laboratory research. Albert Einstein came close to the mark when he wrote:

A hundred times every day I remind myself that my inner and outer life are based on the labors of other men, living and dead, and that I must exert myself in order to give in the same measure as I have received and am still receiving.

—The World as I See It,
Ideas and Opinions (1954)
(trans. Sonja Bargmann)

My gratitude and debts are owed to a larger and more diverse community than that in which Einstein toiled. I was in the fortunate position to write this book because of the individual and collective efforts of a huge community of biologists, including paleontologists, geneticists, embryologists, and evolutionary biologists. While some of the giants preceded my time, most of the discoveries discussed in the book belong to the current generation. I thank the large number of colleagues who provided figures for the book and who have, over a number of years, shared their expertise and ideas.

This is a great job. Everywhere one meets talented and passionate people with a work ethic that would shame most professions. I am especially indebted to the individuals who have worked with and collaborated with me over the past twenty years or so. The creativity and dedication of many students, postdocs, and technicians are responsible for the successes of my laboratory and I have learned far more from them than I ever taught in return. I have also had an unusual amount of freedom in choosing areas to pursue, thanks to the generous financial support of the Howard Hughes Medical Institute, the National Science Foundation, and the Shaw Scholars program of the Milwaukee Foundation.

I had several exceptional mentors who gave me freedom and encouragement in my formative years that catalyzed my growth as a scientist and planted or watered some of the seeds that have flowered here. I thank Simon Silver, Owen Sexton, and James Jones (Washington University in St. Louis); Dr. William DeWolfe (Beth Israel Deaconess Hospital); Dr. B. David Stollar (my Ph.D. advisor), Carlos Sonnenschein, and Ana Soto (all of Tufts University); and Dr. Matthew Scott (my postdoctoral mentor, now at Stanford University) for the exceptional opportunities they gave me and the wisdom they shared.

Finally, I thank my two new mentors in the publishing world without whom this project would not have developed or evolved. My agent, Russ Galen (the four-minute man), has provided terrific advice, astute criticism, and immense encouragement. My editor, Jack Repcheck, sparked the project with his great enthusiasm for Evo Devo and his conviction that this was a story waiting to be told, and he helped to channel my rants throughout the writing process.

Index